Killing Keiko

Killing Keiko

The True Story of *Free Willy's* Return to the Wild

Mark A. Simmons

With Foreword by

Wyland

Callinectes Press

© 2014 Mark A. Simmons
All rights reserved. No part of this book may be reproduced, stored in a retrieval system, or transmitted by any means, electronic, mechanical, photocopying, recording or otherwise, without written permission from the author.

Trade paperback ISBN: 978-0-9960770-1-9
Case-bound ISBN: 978-0-9960770-0-2
E-book ISBN: 978-0-9960770-2-6

Library of Congress Control Number: 2014906071

Photos by Mark A. Simmons, unless otherwise noted.
Cover paintings and chapter art by Wyland
Cover photo of author, by Gary Firstenberg, is used by permission.

Trademarks used herein are for identification only and are used without intent to infringe on the owner's trademarks or other proprietary rights.

Callinectes Press
Orlando, Florida
www.killingkeiko.com

Dedication

This book is dedicated to the millions of adults who, as children, broke their piggy banks to free a whale.

Contents

Foreword	ix
Preface	xv
Introduction	19
1. Getting the Call	31
2. Meeting Keiko and the Release Team	45
3. The Enemy Within	72
4. The Plan for Release—Phase I	87
5. Eternal Daylight	117
6. The Surge	142
7. Phase II, Klettsvik Bay	186
8. The Mean Season	208
9. Welcome to the North Atlantic	239
10. First Contact	286
11. The Unraveling	308
12. Limping Home	317
13. Keiko in the Wild	331
14. Opposing Forces	356
Author's Note	384
Bibliography	387
Additional Reading	388
Index	389
About the Author	397

Foreword by Wyland

Many years ago I received a script to review for a new Hollywood movie titled *Free Willy,* a film about the relationship between a boy and a captive orca. *Strange title,* I thought. The filmmaking team at Warner Brothers, including Lauren Shuler Donner, the producer and wife of the famed director Richard Donner, asked me to consider painting the cover art for the film. Apparently, there was a lot riding on this movie. But I had to wonder, "Who was this killer whale?"

As fate would have it, I had a chance to meet Keiko long after the fame of movie stardom had passed. What I discovered was a long way from the glitz and glamour of Hollywood. Here was this famous whale confined to a small shallow pool at the Reino Aventura Amusement Park in Mexico City. You couldn't imagine a worse place for an apex predator like a killer whale to live. At this time the young male orca was sixteen years old and quite large. His massive dorsal fin was bent like many other orcas in captivity and he was grossly underweight. Worst of all he was suffering from a terrible outbreak of papilloma virus on his face, dorsal fin and other parts of his skin. Keiko also had three unique beauty marks on his lower chin, making him distinctive to all other orcas. But with all that, there was still an impressive spirit about Keiko. He seemed to like everyone and was curious about those who he encountered. That, I was to discover, would later be part of his downfall.

The fact is this young orca was literally kidnapped from his family in the cool waters off Iceland. He was abducted to be sold to an amusement park for profit. Killer whales are big business, generating tens of millions of dollars as entertainment attractions. They are often put in artificial holding tanks and small pools to perform for the millions of people paying premium dollars. This is all great for the park's bottom line, but not so great for the whales. While some parks might have been better than others, the industry of catching and exploiting killer whales has taken a heavy toll on these animals. In the '70s, many died in failed attempts to remove them from their tight-knit family pods that often contain grandparents, parents, and young and baby orcas. The truth is that today even with all our science, we still know little about killer whales and the impacts—on the animals and the ocean itself—when they are removed from their families.

The first time I met Keiko at Reino Aventura he swam over to me and looked deep into my eyes. I looked back into his black eyes and felt a connection. It was like looking into the face of highly intelligent animal who although young possessed the wisdom of his ancestors. Now, here he was alone, thousands of miles away from his home and family, at the mercy of his captors, fighting for his very survival. It reminded me of people who are dependent on the kidnappers for food and water and develop Stockholm syndrome. I couldn't help thinking, *How the hell can I help get this amazing creature out of that cesspool?*

The fact is, they were killing Keiko.

The hot scalding sun and pollution in Mexico City, along with numerous other factors, including malnutrition, were breaking down his immune system. The toxic environment had contributed to the massive papilloma disease affecting a large portion of his sensitive skin.

After the movie, it seemed that everyone wanted to help this poor young killer whale. Millions of dollars poured in from people around the United States and the world. And, like the movie, everyone wanted to free "Willy." The problem was this was no movie. This whale was slowly dying in one of the most polluted cities on the planet. I was invited to meet with Oscar, the owner of Reino Aventura, and Alejandro, who was in charge of the park. It was 1994, and I was planning a West Coast tour to paint a series of ocean murals from Anchorage, Alaska, to Mexico City. In fact, I was invited to paint a mural in Mexico City alongside the great Mexican muralist masters David Siqueiros, Diego Rivera, and Jose Orozco at the university science museum and college. This was considered a great honor because to my knowledge no other foreigner had received such an invitation. The Reino Aventura staff asked if I would paint a mural featuring their star—now controversial—whale Keiko.

At that time no one could figure out how to get the park to release Keiko. I offered to paint a giant public mural featuring Keiko swimming free with his Icelandic pod on the entry of the park. I would paint the mural in exchange for Keiko's release in one year. Oscar and Alejandro agreed. I also had planned one of the murals in Newport, Oregon, at the new aquarium there. A woman named Phyllis Bell from the Newport Oregon aquarium asked me to paint a tiny parking lot wall as part of my mural tour. Maybe, I thought, the climate around Newport could make the aquarium a good potential future home for Keiko. Any place was better than Mexico City.

I told Phyllis Bell that unfortunately the wall at the Newport Aquarium was too small for my life-size murals. But Keiko was still on my mind, so I told her about my idea of trying to find a good place to rehab this amazing animal. She appeared interested but seemed to be holding back. I thought she might have been mad about me not accepting the small parking lot mural for my West Coast tour of murals. The wall at the aquarium just didn't work for the mural project. Instead, I opted for a wall in the town of Newport. Dr. Bruce Mate, a well-known marine biologist spearheaded the project, and even joined me on the scaffolding painting a local whale named Scarback, a gray whale that had a huge scar on his back due to a boat collision. Scarback was the perfect choice to tell the story of the gray whales that migrated along the Oregon coast, and the wall for the image was the perfect size. In the end, the mural was enthusiastically received by the town. At the official dedication ceremony, I somehow ended up standing next to Phyllis Bell and once again told her that my goal was to help free Keiko from the small park in Mexico City. I even stated directly that it would be great to bring Keiko to the Oregon Coast Aquarium.

The West Coast mural painting tour finally made its way down south of the border to Mexico City. Again, I had a chance to visit Keiko and actually got to swim with him for three days. I try not to personify intelligent animals like orcas, but I have to tell you this whale was special. With all that had happened to Keiko, he continued to have a beautiful spirit. He was playful and gentle with everyone. I spent a great deal of time with him, hoping he would one day be united with his family off the coast of Iceland. The first step was to get Reino Aventura to release him. I continued to push for the release by painting one of my murals at the entry to the park featuring a life-size Keiko swimming free. The owners were thrilled and agreed to release Keiko within six months of the completion of the mural. Alejandro even invited me to be with Keiko when he left Mexico City on the UPS plane. *What a great honor*, I thought. *They would inform me when he was ready to go.*

Keiko was to leave Mexico City and rehabilitate at the Newport Coast Aquarium in the coming months. *Great,* I thought. Everything was happening the way I prayed. When it was time for Keiko to leave, I contacted Reino Aventura about next steps and they told me I had to talk to Phyllis Bell at the Oregon Coast Aquarium. I called her but she would not give me any information. She told me I would have to talk to David Phillips from the Earth Island Institute, whom I called the next day. I was told that unless I donated one hundred thousand dollars to the Earth Island Institute, I would never get close to Keiko. David asked me what I contributed to Keiko that would require any further involvement with me. I explained, quite matter-of-factly, that I painted a million dollar mural to negotiate Keiko's release. Finally, David tipped his hand. He said if I flew with Keiko from Mexico City to the Oregon Coast Aquarium, then Keiko and I would receive all the media. He told me point-blank that he wanted the media for Earth Island Institute. I told him that it sounded a lot like extortion. Finally, I said, "No thanks," and hung up.

The next time I saw Keiko was many years later when I was visiting my older brother Steve in Portland. I decided to go incognito to the aquarium to see my old friend. As I approached, Keiko saw me and swam straight at me, pressing his eye against the glass wall of his tank. He seemed to remember me from years ago. He followed me up and down the glass, making sounds. After two hours, I decided to leave him, content in knowing that he seemed to be recovering nicely.

I followed the Keiko story, like many, wondering what would ever become of this amazing animal that had touched so many. Tens of millions of dollars poured in to help Keiko over the years. Kids were giving pennies, nickels, and quarters, anything they had to help free "Willy." Early on people like David Phillips seemed to grab whatever money they could get from Keiko. Others also saw dollar signs and jumped in to get a piece of Keiko. It seems like all the wrong people were connected to this animal that had been abused from the moment he was taken. Where did the tens of millions of dollars go? All that money that poured in appeared to have gone to the wrong people: the people who couldn't care less about Keiko's long-term health and well-being.

I should add something here. There were many great people who dedicated their life to Keiko. Unfortunately, in the end, when the money ran out, so did the support from many of Keiko's purported friends. But that should be a footnote in this story. Killer whales are wild animals that deserve to live their lives as nature intended. They are part of the web of life that keep the ocean healthy and in balance. But they face a multitude of threats, most of them caused by man. Capturing animals to preserve the species might someday be the only choice we have left, but for now no animal should be captured purely for entertainment. In our best zoos, we have successful breeding programs in place. Let's take care of the ones in captivity that have no hope of being released into the wild and, more importantly, let's make sure we take care of their ocean habitat before zoos are the only place left they can survive. The lessons of Keiko and experience have proven over and over again that, as a society, we care enough to do something about it. We just have to use that knowledge and experience to do the right thing.

—Wyland
Islamorada, Florida, 2014

Preface

This story, the story of Keiko's reintroduction to the wild, has been in my head, plaguing me constantly, for the last decade. I began writing the manuscript late in 2000, immediately following my departure from the project. Since then, I have attacked this book many times over the years with elaborate choreographed assaults. My attempts have resulted in various author-like voices, none of which was true to me or Keiko and those who sacrificed much for the love of a whale. Finally, my wife, Alyssa, who lived this experience and adventure with me directed me to write the book as if I was documenting Keiko's story for my daughters. This voice, told to open hearts and minds, was much more natural for me. I hope it is for you, the reader, as well.

Keiko's story would never have seen the light of day had it not been for the tireless support of a few notable family members, friends and colleagues. Lisa Lauf Rooper, you are my muse. Without your inspiration I would have cowered away from this task long ago. Andy Schleis, my humble thanks for a kick in the pants when I most needed it. To my beautiful, intelligent wife—my "wingwoman"—the Bible says that through God all things are possible; you have stubbornly championed His word in ways that I cannot describe.

My most humble thanks to the many lifelong compatriots from the Keiko Release Project for their feedback, many broken hours dredging up tender details and for keeping me on the straight and narrow. Jeff Foster, Michael Parks, Jim Horton, Guðmundur

"Gummi" Eyjólfsson, Kelly Reed Gray, Thomas Sanders, Mark Trimm, Sigurlína Guðjónsdóttir, Smári Harðarson, Ingunn Björk Sigurðardóttir Bjartmarz, Sveinbjörn Guðmundsson, and Tracy Karmuza McLay. You each have my undying gratitude.

To Stephen McCulloch, your encouragement during the last many years, and your passion for our shared field has been nothing short of awe inspiring. Further, your wise and timely introduction to Wyland proved a remarkable foresight. Wyland's publishing experience kept this story from certain disaster at the hands of the wrong editors. His personal investment in Keiko's life and lessons lent itself to a collaboration uncomplicated by senseless controversy. We both share the same hope (and urgency) for a more collaborative future in marine life conservation.

Lastly, I thank my dear friend Robin B. Friday Sr. for his place in this journey. You are my brother in Christ and will always have my respect and admiration.

Killing Keiko

Introduction

Let us take things as we find them. Let us not attempt to distort them into what they are not. We cannot make facts. All our wishing cannot change them. We must use them.
                                                John Henry Cardinal Newman

Before meeting Keiko, I knew him well.

There are those who would have you believe that killer whales at marine parks are somehow different than their wild brethren; that the whales in zoological settings are crazed by years in "prison." In more than twenty-seven years working with and around killer whales in parks and the open ocean, I have never seen one ounce of evidence to support such a statement. What I have seen is that these amazing animals adapt well to almost any change. They are top predators, and among their many talents is the ability to thrive in many environments.

The playful interactions I have witnessed among killer whales in the wild are identical to the play I've watched thousands of times among whales in marine parks. Breeding and social behavior is identical, even cooperative hunting is often witnessed with zoological whales when they bait and trap the occasional seagull. The primary difference is that wild whales spend much more time engaged in hunting and traveling (primarily to hunt or follow migrating fish) and less in play and social behavior. Although killer whales certainly engage in different behaviors depending upon their surroundings, the animals I have witnessed all share one common unmistakable characteristic: the disposition of the ocean's top predator.

Killer whales are like the lone wolf in a busy dog park. They carry an air of superiority over all others around them and are the masters of all they survey. They have the temperament of calm and benevolent dictators descended from a royal bloodline. This disposition is remarkably consistent among every killer whale I've ever witnessed. Killer whales do not act crazy, for it is beneath them regardless of their circumstances.

However, as it is with every living thing on this planet, an individual animal's behavior (how it acts and what motivates those actions) is influenced by a multitude of variables. This precept frames perhaps the most vital dynamic when introducing an animal to significant change. As the science of behavior constitutes the foundation by which an animal's behavior can be successfully altered; the species, the individual animal's traits and learning history from birth thus comprise the medium. All past experiences shape the likelihood of how future behavior will occur. It is no wonder that the science of behavior along with Keiko's past in the company of man would play the key roles in his return to the wild.

Killer whales are not dolphins. I've heard too many times, "Killer whales are just dolphins; they are the largest member of the dolphin family *(delphinidae)*." Rubbish! These are the words of a taxonomist. They may be scientifically categorized in the same family as dolphins; however, behaviorally killer whales are no more like dolphins than rottweilers are like retrievers. Anyone telling you different is likely to get you maimed or killed.

There is no limit to the number of self-proclaimed experts out there or those touting "Dr. This or That" and "Suzie Q, Ph.D." in killer whale biology, ecology, and a few other "ologies." I am not overly impressed. One can study something from a distance and write all the papers he or she likes, but the more distal and passive methods of study contribute little in the way of truly understanding an animal like a killer whale. Put yourself front and center in their watery world, neck deep in glorious vulnerability . . . only then do you begin to understand how they think and what they are made of; all preexisting assumptions are quickly cast asunder. There, face to face with a killer whale, adrenaline fuels a crisp mind free from distraction and with laser-like focus. In those moments, close attention is paid to what is real.

After working with many different animals in a zoological setting, I can think of no other animal that compares with a killer whale. Thinking back about working with polar bears, walrus, sea lions, various species of dolphins, false killer whales, canids and birds, there always seems to be something profound missing in those interactions. No doubt these latter relationships are special and wonderful, but that one extra indescribable ingredient with killer whales makes the cake rise just a little higher, taste just a little richer.

Let me be clear, never is the illusion of control more apparent than when working *with* a killer whale. It means working together; there is no dominion over the animal. It means depending on each other and depending on a strong and well-established relationship, a working relationship that earns the animal's attention and trust over time. It is a partnership, a cooperative effort, not a trainer leading a whale and certainly not that of a boss dictating to an employee. The partnership is 50/50 and each fulfills a distinctive role in that relationship. The trainer must do his or her part to protect a hard-earned trust and bond utilizing every ounce of intellect, skill and ability to read that animal and provide a clear, consistent atmosphere of learning.

Killer whales are constantly communicating to their partners in a training environment. It is up to the trainer to learn the language of killer whales and the idiosyncrasies of the particular animal. The basis for that communication is the science of behavior and the art of applying behavioral modification. Simply put, this means that *environment, cues,* and *consequences* shape the way an animal behaves and the choices it makes. This principle is not exclusive to zoological settings; rather it is a hallmark of nature, existing in every setting. Nor is it exclusive to other animal species; it holds true for humans as well. Influence on behavior is like gravity, always present. Behavior is always in a state of change and always able to be influenced. At any given time, there is a vast array of forces at work shaping the way an animal (or a person) behaves.

Understanding the principles that shape behavior is only the beginning step in the process of introducing change. There must also be an understanding of the particular species' genetic history, the individual animal's learning history since birth, its physical capabilities and limitations, its environment (past and present), its social setting and where it fits within the social hierarchy.

To effectively manage how an animal's behavior is shaped, variables that can be controlled must be identified and managed proactively. Environmental conditions that set the stage for certain responses, other animals, human activity, sounds, the frequency of and the predictability of every stimulus perceived by the animal; these are but a few of the variables within one's sphere of influence. In the case of Keiko (Keh-ee-koh) and his reintroduction to the wild, that sphere of influence was sophisticated and many layered. There were many potential factors outside this sphere of influence for which there was no control. Nonetheless, there had to be a masterful awareness of all potential factors at work in Keiko's environment.

Potential influences outside of our control can be anything and happen at any time: they are like uninvited guests who crash a party. A good host will pick up on these situations and gracefully turn them to his or her advantage. The complexity of Keiko's release was on an order of magnitude light-years beyond what Hollywood portrayed in the film *Free Willy*. In order to offer him the best chance of success, the application of learning principles and the constant analysis of all areas of influence had to be fluid and unrehearsed. There was much at work that pitted the odds against Keiko's success; however, when all the elements shaping behavior are managed appropriately, anything is possible.

I spent the most formative years of my young adult life working alongside killer whales in a zoological setting. There is no complete manual on how to work with a killer whale that I know of, but if there were, it would, among many other things, advise new trainers that reacting with fear or intimidation will get them killed. This is not to be confused with being stupid or taking unnecessary risks, but one cannot be timid with a killer whale. Being a top predator, they are not prone to "small talk." They carry an impressive array of sensory capabilities, and they are deliberate in everything they do. You cannot sneak up on them. They are 100 percent more aware of their environment than we are.

The most remarkable illustration of a working relationship with a killer whale I've seen (outside of the real thing) is depicted in the movie *How to Train Your Dragon*. The connection and delicate trust portrayed between the lead character Hiccup and the Night Fury dragon masterfully illustrate many uncanny parallels found in the bonds between trainer and killer whale.

A trainer has nothing that will impress a killer whale, and a killer whale has no burning desire to impress that trainer or earn his attention. If the animal works with his trainers, it is because it chooses to do so, and that choice is most definitely based on relationship; on the excitement and challenge its caregivers can offer. One cannot motivate a killer whale with food for long. Leveraging food to motivate animals is an antiquated method used by a few marine mammal trainers in the 1950s and '60s. Killer whales are immune to this approach. With a formidable layer of blubber, they can easily go without food for several weeks. I have witnessed killer whales offered salmon (a delicacy) take the tendered fish only to spit it, almost perfectly fileted, back in the lap of the unwelcome trainer. I've seen countless times when a whale has refused to work with a certain trainer, only to meet the favored next with unbounded energy and enthusiasm. Food is not a compelling motivator. Make note of this important lesson; it is a critical component to understanding Keiko's journey back to the wild.

To understand Keiko, the effort to return him to the wild, and the project's outcome, three defining factors stand out above all others: 1. recognizing how learning occurs and behavior is influenced; 2. knowledge of what it means to work with a killer whale in a training environment; and 3. awareness of the distinctive traits of a killer whale. They are the only means by which decisions made throughout Keiko's journey can be weighed against their impact on his behavior, his choices and the final outcome.

Beyond a story about the invisible forces of nature and learning taking place on the high seas of the North Atlantic, this is more so the tale of one killer whale and his notorious journey to freedom. It is about a journey that spans four decades, encompasses the zoological and animal rights communities and epitomizes the evolution of public appreciation for the killer whale. Keiko's story begins long before Hollywood uncovered an icon in *Free Willy,* or before children around the world recognized compassion.

In order to understand the profound breadth of Keiko's journey, it is necessary to begin with an understanding of the industry that created him and, ultimately, how philanthropists urged Keiko to follow in the footsteps of his fictitious counterpart, "Willy."

Whale Killer

Fifty years ago, the general perception of killer whales was that of maniac predators roaming the seas and ravaging anything in their path. Fisherman hated them for stealing bounty from their nets. Whalers hated them for devouring their catch as it was towed alongside their ships. The military used them for target practice from ships and in aerial simulations. This loathing, along with the public's false impressions of the ocean's top predator, vanished almost overnight following a series of unexpected events that unfolded in the mid 1960s.

In 1964 a male killer whale named Moby Doll was harpooned by an expedition commissioned by the Vancouver Aquarium. They intended to kill the whale for skeletal fabrication and subsequent display in the aquarium. The whale was harpooned but did not die. Instead Moby Doll lived for eighty-seven days in a temporary Vancouver-based sea pen. In that short time, he became an international celebrity and attracted scientists and the public alike.

In 1965 yet another encounter with a killer whale excited public attention. After a male killer whale had been caught in a gill net near Namu, British Columbia, aquarium owner Edward Griffin towed him over 450 miles in a makeshift sea pen to the Seattle Public Aquarium. Songs were written, and a movie titled *Namu, the Killer Whale* was made. Namu died there after only eleven months, believed to have succumbed to an infection from poor water quality. However, during this time he developed a relationship of sorts with Griffin and became the first live orca to perform in front of the public.

Seemingly overnight, an industry was born. Killer whale collections for the purpose of public display began in Puget Sound. But by 1976, due to opposition from environmental factions and public sentiment promulgated by the death of five whales, collections in the Northwest were halted. Thus Icelandic waters became the next frontier for killer whale collection boasting larger populations, a capable shipping channel, and the indifference of a whaling nation.

The ideal killer whale candidate for collection was usually between two and three years of age. But determining age was less than precise. Many animals much younger than two years old were taken. Following an initial acclimation period, graduate whales were transported to zoos and aquaria around the world. By the 1980s, "themed" animal parks were located on virtually every continent. Due in large part to unparalleled intimate exposure, the general public quickly gained an insatiable fascination with the killer whale.

Science was only just beginning to understand the extraordinary learning ability of the bottlenose dolphin when this mysterious and beautiful cousin took center stage, filling our hearts and exciting our minds. The striking and bold coloration of the killer whale became its trademark. Crisp mirrorlike black contrasted by the milky-white underside made it appear as living, breathing art. Likewise, the distinguishing combination of strength, beauty and social complexity appealed to a wide audience. The top predator of the ocean was supreme in disposition, mysteriously elusive at sea and the top of its class in the animal training environment. For most, it took little effort to fall in love with or be awestruck by *Orcinus orca*.

In the late '60s renewed interest in migration routes, populations and social behavior led various individuals and organizations to study killer whales in the wild. In efforts to collect data, many researchers devoted their lives to the task, spending tiresome hours onboard research vessels with few accommodations. Every aspect of the killer whale was an exercise in stretching the imagination. What had been recently believed to be a ferocious man-eating killer was now the focus of unlimited study and public interest. At the same time, knowledge of whales was growing in leaps and bounds due to the relative ease with which researchers could observe the whale's behavior beneath the water's surface in zoological settings. Trainers and caretakers gained an unequaled respect for the killer whale's aptitude. Many of today's advancements in behavioral conditioning and the application of positive reinforcement techniques in the animal training field were originally implemented and streamlined through work with the ocean's top predator.

Keiko's Collection

According to the "Reintroduction Protocols," the formal document submitted to the Icelandic government from which Keiko's release permit was granted:

> The killer whale (*Orcinus orca*) "Keiko" was captured off the coast of Iceland in 1978 at the estimated age of two years. Following two years at a temporary housing facility in Iceland, Keiko was transported to Niagara Falls Aquarium and maintained there for a period of six years. In 1986 the subject was transported to Reino Aventura Amusement Park in Mexico City, Mexico. While maintained in Mexico City, the subject's health progressively deteriorated due to inadequate environmental conditions.

Immediately following collection, Keiko was taken to a coastal sea pen-type facility in Iceland. He was maintained there for an extended period of time, as was common for newly placed animals that would eventually go to permanent facilities. This period of acclimation ensured that individual animals were healthy, eating, and had successfully adapted to the initial change.

The first move from Reykjavik, Iceland, to their new home was the toughest transition for the whales. It involved yet another change of environment or acclimation period for the whales, this time with a completely unfamiliar social group.

During Keiko's almost six years in Marineland, Niagara, he was socially ostracized, physically displaced (picked on) and constituted the bottom of the hierarchy within his new social group. There were an estimated six killer whales in total at Marineland during Keiko's stay in Canada.

After being sold and moved to Reino Aventura Park in Mexico, he developed a skin disease called cutaneous papillomatosis caused by a novel papillomavirus and associated with immune suppression (which is a potentially contagious disease to other whales). The condition formed an unsightly cauliflower-like growth of the skin where affected. Keiko quickly gained a stigma as the ugly duckling among those who cared for him. Even so, he was the marquee attraction at Reino Aventura for nearly eight years before achieving Hollywood stardom.

Free Willy

In 1993 Keiko became the star of the Warner Bros. blockbuster movie *Free Willy* and, as a result, inarguably the most famous killer whale in history. The movie depicted a killer whale (Willy) as languishing, neglected in a small pool at a theme park. In the movie, Willy is befriended by a lone boy and eventually spirited away back to the wild where Willy swims off into the sunset and lives happily ever after (including Hollywood's production of three sequels).

After the movie, the nonprofit animal rights organization Earth Island Institute began lobbying for a real-life release program for Keiko, intending to have him follow in the footsteps of his fictitious counterpart. In 1994 Warner Bros. contributed $4 million to the movement, and Earth Island Institute formed the Free Willy/Keiko

Foundation (FWKF) to spearhead the release effort. Reino Aventura, under public pressure, donated Keiko to the Free Willy/Keiko Foundation. In 1996, as the first step of a program to return Keiko to the wild, he was transported to a newly constructed facility at Oregon Coast Aquarium in Newport, Oregon. The facility cost nearly $8 million to build.

According to the formal "Reintroduction Protocols" from the Keiko Release Project permit:

> In January 1996 Keiko was transferred from Mexico City to the Oregon Coast Aquarium (OCA) in Newport, Oregon. Objectives for this period were to improve general health and quality of life by providing a high-quality environment and structured rehabilitation program and to provide for appropriate public display opportunities. Following two and a half years of successful medical rehabilitation, on 9 September 1998 Keiko was transported from Newport to an open-water bay pen facility in Vestmannaeyjar, Iceland. The transport was conducted, pursuant to the transfer provisions of section 104(c)(2) of the Marine Mammal Protection Act, under an authorized National Marine Fisheries Service (NMFS) public display permit and an export permit issued under the Convention on International Trade in Endangered Species of Flora and Fauna. Iceland was chosen as the site for potential reintroduction due to the fact that he was originally captured in Icelandic waters. As part of the transfer operation, a public display program was established and carried out.

From September 1998 to February 1999 in his Icelandic bay pen, Keiko did little more than continue to gain weight. Theories and ideas about how to move forward with the release were abundant and diverse, but mostly the Free Willy/Keiko Foundation expected Keiko to take the initiative toward his freedom once in native waters. Some expected him to call to his brethren; some envisioned Keiko's mother swimming up to the bay pen and coaxing him to follow; and still others imagined Keiko would be fattened up and taken to sea to be dropped off with wild whales, where he would obediently swim off into the sunset.

These were not only the ideas of the children who broke their piggy banks to contribute to Keiko's release; these were the machinations of the board of directors, the founders of Earth Island Institute and the head veterinarian of the FWKF. Despite the numerous and idealized visions of release, no concrete plan of reintroduction was ever established beyond his relocation to Newport and later transfer to Iceland. The FWKF literally did not know what to do next. Keiko adjusted to the climate and waters of Klettsvik Bay, Iceland, where his bay pen was located. Beyond his weight gain and developing an unhealthy attachment to a large Boomer Ball (a three-foot diameter hard plastic ball, his sole companion at the time), Keiko achieved little in his preparation for the wild during his first five months in Iceland. The program managers and the board of the FWKF were at an impasse. They had inherited an iconic whale and an extremely expensive operation to maintain, but no forward progress materialized.

To begin to understand the challenges facing Keiko and the team of people charged with his reintroduction, any comparison to Hollywood's version of *Free Willy* must be cast aside. Releasing an animal, any animal that has spent considerable time in the care of man, is a complex process to say the least. In the case of Keiko, it was analogous to putting the first man on the moon. If a candidate for release were imagined, many experts agreed, it would not be Keiko. An adult male killer whale, dependent on social acceptance for his survival, was the least likely to be accepted by his wild counterparts. Chances were other males would view him as a competitor. But this aspect was only the tip of the iceberg challenging Keiko's survival. After nearly twenty years in the care of man, much greater threats lurked undetected in Keiko's chances of survival.

100 miles

Keflavik
International Airport

Reykjavik

Vestmanneayjar

ICELAND

NORTH ATLANTIC OCEAN

N

Getting the Call

June 15, 2000–1921 hours

Communication between tracking helicopter (call-sign *Zero-Nine-Zulu*) and the *Draupnir*, Keiko's walk-boat:

Zero-Nine-Zulu: "Contact. We have positive sighting . . . advise heading east-northeast, repeatedly circling then continuing course. Be advised fuel is short . . . heading back to base."

"Copy that, *Zero-Nine-Zulu*, we are closing on your location at twenty-six knots. Tracking equipment onboard. *Draupnir* out."

Moments later, onboard *Draupnir* via ship-to-shore phone message: *Draupnir:* "Hello?"

From base: "Hi, Robin, it's Charles . . . I have Jeff and Lanny here."

"Okay," Robin replied, adding under his breath, "This should be interesting."

Lanny spoke first: "What are you guys doing?"

"We're going to get this whale back . . ."

"Why?" Lanny challenged. In a tone of complete condescension, he stepped up the attack, "That's against the protocols that we set in place—just because he didn't go with those whales doesn't mean he won't eventually go with other whales. He may be heading home, and you guys are calling him back. You talk about going against the protocols; our protocols were always that if he decided to go off on his own to let him go!"

Knowing all too well the conflict of interest behind Lanny's motivation, Robin could scarcely contain his anger. He muttered,

"The bastard just wants his success fee." In the midst of this desperate mess, Robin had no patience for Lanny's outburst and made no attempt to conceal it.

"Lanny, you're wrong! That was never the protocol! From day one, in our first meeting at the hostel, we all agreed successful reintroduction would be only in the case of his successful integration with other killer whales. Right now he's alone, he's traumatized, confused, and he doesn't know where he's going!"

Unwilling to back down, Lanny pressed, "But we said we would not immediately intervene—that we would allow time to observe his disposition and then make a decision whether to recall him to the boat."

"We are already approaching fifteen miles from the island," Robin snapped back. "If we allow him to go any further away, we will be too far from our base of operations to be able to monitor his disposition and/or intervene should that become necessary. In my opinion, the bottom line is that he is not successfully integrated . . . that the initial introduction was a fiasco and Keiko is simply running scared! My intention at this point is to find him—make an observation—recall him—and bring him back to Vestmannaeyjar. If the final decision is to allow him to go off on his own then that decision can be made after we bring him back—and that would be a decision that you gentlemen would have to make on your own . . . without me."

Charles interjected before Lanny could respond. In a calm reassuring voice he took the reins: "Robin, we have talked to members of the board . . . advised them of the situation. They want us to make the decision of what needs to be done. I think Jeff and I agree that you should bring him back—once you locate him—you should bring him back."

Lanny wouldn't let it go. "Well, I think it's wrong, and I disagree."

Keflavik International

I sat slouched in the main airport terminal, both hands stuffed in my jeans pockets while watching the concerned look on everyone's faces. It seemed surreal in so many ways, down to the stage set at this our final departure. Shell-shocked and confused, the crowd surrounding me looked as I felt. Glancing overhead, I could just make out the last few back and forth movements of the life-size biplane suspended from the airport superstructure.

I had never been through an earthquake before; yet, when it hit, somehow I immediately knew that is was an earthquake. It's not difficult really when the floor moves under foot at the same time the ceiling is swaying in the opposite direction, confounding the senses. Not much other than an earthquake could rattle your whole world like that. Well, that . . . and Keiko. My thoughts vacillated between the swaying fixtures overhead and the prior day's fateful exchange onboard the *Draupnir*.

The earthquake, rated 6.5 on the Richter scale, had provided a dramatic exclamation point to our final departure from Iceland and the Keiko Release Project. I didn't know if the Nordic gods were punishing me or someone else back on the island, but there was plenty of time to think it through. My departure was unplanned, and as a result, I was summarily placed on the standby passenger list. To add insult to injury, the earthquake had collapsed a fuel cell somewhere on the tarmac. The airport officials suspended all flights for a few hours while they checked things out. I had nothing but time.

Airports are known as good venues for people watching. Keflavik International (kef-la-vik), a hub between Europe and the Americas, was among the best. I tried repeatedly to distract myself with the commotion of people around me, but to no avail. The weight of it all was crushing, *How the hell did I get here? What am I doing in Keflavik? I shouldn't be here, not today anyway.* I knew how it happened, and it all seemed logical and sequential in my head, but I still couldn't understand it. I swallowed voluntarily, testing the lump in my throat. One minute everything made sense and the next it made no sense. My only consolation: the confidence I was in good company. Two seats down to my right, Robin was lost in his own thoughts.

Friday, Like the Day

The path leading up to this day was nearly fifteen years long. Along that path, the one constant I could always rely on was Robin Friday. He and I shared different views on many things, and we constantly analyzed and debated everything. At times our method was exhausting, some might say obsessive, but it worked. Whenever we came to agreement on a topic, you could be sure it was thoroughly vetted and never rash; especially not where it concerned Keiko.

Robin had been the curator of animal training at SeaWorld of Florida and my boss during my last few years as a killer whale trainer there. A natural leader, he was a savant at working with animals; he had a certain way around animals that was indeed rare. In his silent hands-off way, Robin also stood out with individuals; secure enough in his leadership to allow people to push beyond their status quo even when that meant making the occasional mistake. He knew that making mistakes and learning from them was essential for progressing in life and work.

Robin has never been one for many words; the quiet type, but always exuding competence. When he speaks, it is only because he has something worthwhile to say. He was then and is still a handsome man, mostly graying hair with that Marlboro Man brand of ruggedness about him. I'm not the worst person to look at, but the never-ending giddiness of females whenever Robin was around made me feel like climbing the bell tower in Notre Dame. To most, Robin is humbly disarming. To all, he is exactly the kind of person to have alongside when facing the most trying times.

Robin was a master of many things, but his specialty lay principally in the area of animal husbandry. In the zoological field, this is the science and art of ensuring animals in human care are healthy, socially well-adjusted and happy. This is not a nine-to-five profession. Robin's chosen field involving marine mammal rescue, rehabilitation and the occasional animal transport placed round-the-clock demands on his personal time.

While he wasn't a trained veterinarian, Robin often knew more about marine mammal care and practical application than many of the vets on staff. He also traveled extensively on behalf of the company. In fact, when it came to marine mammal rescue and rehabilitation, he had rescued nearly 300 animals at this point in his career and lost many a night's sleep in the process, caring for fragile survivors around the clock. Animal rescue was, and to this day remains, his passion. When another facility or a government came to SeaWorld for expert help, more often than not, Robin led or was part of the team of responders. He would never say so himself, but Robin Friday is one of the most well-known and liked professionals in the marine mammal community.

My experience with animals was focused on behavioral science: the application of behavioral modification commonly known as "animal training." Having realized a dream that transported me from Virginia to Orlando, I began working with marine mammals at the ripe age of eighteen. I spent the following ten years of my life at SeaWorld in some of the most fascinating and unbelievable circumstances, dedicated to understanding and shaping killer whale behavior.

By the time I left SeaWorld in 1996, I had managed Shamu Stadium, represented SeaWorld in British Columbia on the acquisition of three killer whales from Sealand of the Pacific and participated on numerous marine mammal rescues involving dolphins, manatees and the occasional pygmy sperm whale. My time at SeaWorld was life changing. Nothing compares to an intimate working relationship with animals, especially when those animals weigh in at five tons and are sharper than many of the people that work alongside them.

When Robin decided to leave SeaWorld to take a general manger position at another marine life park, I remember telling him, "Never hesitate to call on me, I'll gladly follow wherever you go." Less than a year later, he did precisely that, and we have since spent our careers working together.

After leaving SeaWorld and finishing my business degree, Robin and I formed a professional partnership, creating a zoological consulting business. We both tend toward altruistic ideals and are passionate about our trade. As a by-product, our business objectives were equally benevolent and far-reaching. Our goal was to cross traditional boundary lines with our new organization and in so doing, to share a considerable arsenal of knowledge and experience to the betterment of animals and wildlife management. We were both blessed to have graduated from SeaWorld, the "Harvard" of the marine mammal zoological world, and we intended to spread this wealth of knowledge. Our focus was not solely public display facilities; we would seek out any case where the care of marine mammals was deficient and, of course, where the proprietor or government agency was accepting of outside help. That last criteria proved to be the toughest.

Even so, there were enough projects to keep our small organization busy throughout the beginning of 1999. Much of our time was spent networking, which ultimately gave rise to our contact with the Keiko Release Project. Robin had an extensive list of close contacts in the zoological field, and his professional reputation opened many doors. In particular, it was an antiquated relationship with Keiko's head veterinarian that opened the door to meeting the most famous killer whale in history.

The day Robin received the call about Keiko I was in Colorado visiting a close friend. Robin was requested to visit the Keiko release operations in Iceland and explore the possibility of working on the project. I remember immediately thinking, *He's freaking crazy*. Within the professional zoological world, the Keiko project was highly controversial, and there was no doubt in my mind, our involvement would be a risk to our professional futures.

The issue was not about releasing an animal to the wild; we had worked on release programs before. But those release programs involved only stranded or distressed wild animals, animals that had not been in the care of man for very long. Releasing a zoological animal that had been in the care of man for decades was a completely different beast altogether.

Even in the field of marine mammal strandings, there is much controversy regarding the effects of being in the care of man, if only for a short time. During rehabilitation the unavoidable association with humans impacts the animal's ability to survive once released back to the wild. In fact, the success rate when releasing a rehab animal is not good, even though the animal might have lived the vast majority of its life in the wild. This of course is dependent on many factors; however, prolonged behavioral conditioning in association with humans is often an overlooked and underestimated force, capable of casting an all-powerful veto over every other advantage toward survival.

It is not uncommon for dolphins that have been rescued, rehabilitated, and then released to exhibit nuisance behavior, following boats and seeking handouts; this after only a brief association with human contact. By U.S. standards, any animal that is rescued and in the care of man for even six months must go through a comprehensive rehabilitation process in order to be approved for release. That process involves avoiding counterproductive associations during release and the systematic removal of dependencies before the animal is returned to the wild. Releasing a longstanding captive adult male killer whale to the wild? The U.S. regulatory agencies would never allow such a preposterous release program from our waters, and believe me, there are many good reasons why not.

I brushed off the idea of our participation in Keiko's release and reminded Robin of the implications. But as the day wore on, the concept became increasingly more fascinating and very difficult to idly dismiss. It actually seemed a perfect match for our business mission and background. It was almost as if everything we had done in our professional careers had been a primer for this project, with each experience and exposure culminating to prepare us for an undertaking of this magnitude. Or at least that's how the idea of it began to resonate.

As one who had spent my career shaping animal behavior, the Keiko Release Project represented the ultimate challenge: To train an animal for every conceivable skill required to survive the wild; to remove decades of conditioned dependency; and to eliminate or replace the human-animal bond. The idea was overwhelming really, but also stirring. This was an opportunity to apply every ounce of available science in behavior modification with no margin for error. It would require extensive marine logistical capabilities, input from trusted behaviorists, and span an unknown quantity of time. Without a doubt, it would depend heavily on deep pockets to finance all of it.

Money didn't seem to be a problem. Based on what the project managers had shared with Robin, everything needed was in place—from specialized marine equipment and the temporary bay pen facility to staff support and extensive monitoring equipment, and especially, the financial commitment to bring it all together for as long as it would take. (The pen itself was a temporary housing site intended to acclimate Keiko to the varying temperatures, sounds and currents of the natural seawater and open environment.) The only thing the Keiko project team lacked was the experience to create and carry out a reintroduction plan that placed an intense focus on Keiko's learning history. Sleep never came that night; my mind went on autopilot, already hard at work on the prospect.

Following a week of debate on the issues surrounding the infamous release, Robin and I agreed that he should at least visit the project and find out more. By mid-February 1999, he accepted the invitation. The Free Willy/Keiko Foundation (FWKF) flew Robin to Iceland to meet with project leads, evaluate Keiko and the people closest to the reintroduction effort, and learn as much about the project and the people as seven short days would allow.

Testing the Waters

Our primary concern at that time was deciphering the organization in charge. In other words, who was "behind the curtain," and what were the stated and unstated motives? We needed to know that life and death decisions would be made with Keiko's best interests at heart, even if those decisions eventually conflicted with what the organization had sold to the public. Like any undertaking that costs money, raising it requires a clear goal, marketing and a return on investment. This project was funded by private wealth and children's piggy banks from across America and Europe.

We knew quite clearly what the public and private donors had been spoon-fed, which was nothing short of a convoluted Hollywood version of life. We knew promises had been made and reputations were on the line. What we didn't know was how the people at the helm would respond if Keiko didn't make the cut. After all, the movie *Free Willy* convinced the world that releasing a whale was as easy as plopping it in the open ocean.

Robin and I understood rather well who some of the more colorful characters were on the project's board of directors. These were the quintessential antagonistic animal-rights activists that were famous for shockingly crass and ignorant statements, even outright lies, about zoos, animals and individuals. By their actions or their words, these were not rational people; nor did they advocate moderation or collaboration. Their ilk were notorious for statements such as, "The life of an ant and that of my child should be granted equal consideration," and, "Phasing out the human race will solve every problem on earth, social and environmental."

These specific groups that had come to manage Keiko's release represented the antithesis of the zoological community. It was clear that association with this project would constitute a defection from many respected relationships in our profession. Still, we had vowed to cross these boundaries, not to employ the same tired and hostile political strategies in dealing with them, nor allow such barriers to prevail at the expense of an animal we had the opportunity to help. Throughout Robin's first visit and our distanced communication, we toiled with the makeup of the organizations behind the project, but our exchange always gravitated toward a solution. Even if neither of us would outright admit it yet, we both wanted to tackle the challenges facing this project.

At the heart of the issue, Robin and I felt we were different. We thought, perhaps foolishly and maybe idealistically, that we could be collaborative. To some degree we welcomed the challenge to educate and hopefully bridge the gap in philosophies that had created the Keiko Release Project.

Above all else, our primary concern was that Keiko's best interest would be the priority. Assuming a focus on Keiko and his needs could be verified and the basis for sound decision-making entrusted largely to us, Robin and I were convinced that we could much improve Keiko's chances of success.

Although the usual suspects in the fanatical faction of the animal rights movement were definitely involved in the Keiko Release Project, fortunately they were contained at the board level or stayed on the periphery of the project, well away from Klettsvik Bay, Iceland. Ironically or poetically, people with a zoological background were the frontline running the operation. Surreal in many ways, they had employed the very people they campaigned against in order to facilitate Keiko's release to the wild.

This was true from the start, including Keiko's first journey from Mexico to Newport, Oregon, throughout his care there and onto his transport to Iceland. Keiko was constantly under the supervision of individuals from the zoological community. This was very odd and always stood out to me like a huge elephant in the room; one that no one dared discuss openly.

At the time of our initial contact with the project, the actual personnel attending to Keiko's day-to-day care in Iceland were individuals with whom Robin had worked in the past—namely Jeff Foster and Peter Noah, the two on-site project managers. Jeff and Peter both shared an animal background not unlike that of Robin's and mine. They were not raised in the culture of anti-captivity supporters and were therefore down-to-earth in their approach to the project and us.

Their way of thinking about Keiko's release was based on their experience, not a philosophy. Jeff and Peter's leadership stemmed the radical tide, and to a certain extent, made the project palatable. Ultimately, the people and the perspective emanating from the frontlines with Keiko made our decision for us.

The Proposal

Upon Robin's return to Orlando following his initial visit, Dave and I commenced lengthy and relentless interrogation. Dave was our third in the "three amigos" makeup of our new company. He shared a similar background with me, having worked with killer whales and studied behavioral sciences. It was our job to take what Robin reported—his evaluations and opinions—and put all of this into a reintroduction outline and formal proposal, a plan deeply rooted in behavioral rehabilitation or the systematic reprogramming of Keiko.

Once completed, this proposal was presented to Ocean Futures Society (OFS), a newly formed nonprofit born of a joint alliance between the Jean-Michel Cousteau Institute and the FWKF. OFS was the front line in the management of Keiko's day-to-day needs; however, they answered to the FWKF board on all things "Keiko." Our proposal represented a vast divergence from their previous approach and placed the principal focus of Keiko's release on behavioral modification. It also positioned our company as subcontractor to OFS for implementing the plan.

Apparently OFS found some favor in our outlined plan of reintroduction, but exactly what I do not know. To our dismay, the project's head veterinarian, Dr. Lanny Cornell and the board of directors did not consider that *behavior* had anything to do with Keiko's preparation for release. This shortcoming left a cataclysmic gap in their concept of Keiko's introduction to the wild. Amidst all the management inadequacies apparent thus far, including lack of experience with killer whales and the absence of a structured release plan, this was the Grand Canyon of them all. They blatantly failed to recognize or even consider the impact of Keiko's life over the past two decades and his learning history. Quite simply, they believed that Keiko would "figure it out." Much like the flatbed trailer scene in *Free Willy*, in which Willy was backed into the sea from a boat ramp, the leadership of the FWKF believed Keiko's release was primarily logistics, just getting him to Icelandic waters.

Dr. Cornell had formerly been SeaWorld's head veterinarian in the 1970s and '80s; the only individual early in the project with any killer whale zoological experience worth noting. Lanny, however, was a veterinarian. While trained and studied in marine mammal medicine, he knew very little about shaping behavior. It was a classic case of Maslow's hammer: "If all you have is a hammer, then everything is a nail." Our experiences with Lanny in the past attested to his approach: behavior was treated with drugs, not conditioning, and behavioral science was fool's play. In fact, the project's staff at that time referred to much of Lanny's direction as "voodoo science."

To say Lanny is well known for his loathing of animal trainers would be an understatement. In his previous career at the helm of SeaWorld veterinary care, he was also known as a bit of a bully. Sadly, there had been no shortage of stories recounting his notorious intimidation tactics.

This sordid history left no question in our minds that Dr. Cornell had played an enormous part in dismissing behavior as having any relevance in Keiko's rehabilitation. Early on, we often pondered the apparent discrepancy, *Why then had Lanny been the very person to invite Robin's involvement?* More than a decade had passed since the two had worked together. It's most likely that Lanny was oblivious of Robin's professional evolution, which had brought about a very different attitude on the subject of behavioral science. In fact, Robin had come to consider behavior a critical foundation in the care of marine mammals, most especially one such as Keiko.

Even so, despite the heavy focus on behavior at the heart of our proposal, something about our outlined reintroduction plan provided an opening. As a result, and to our surprise, Robin was requested to return to Iceland in March, this time for a thirty-five-day tour. On this visit he would become actively involved in privileged day-to-day operations, notwithstanding the lack of any formal engagement. We had multiple exchanges during his first two weeks back, mostly concerning Keiko, Robin's observations, changes he made to daily interactions and his overall assessment of the operation.

Although Robin didn't specialize in behavioral science, he was adept at evaluating and recognizing effective and ineffective applications and describing the need for behavioral modification. Over the course of his second tour in Iceland, he was able, through example and education, to convince the decision makers that additional behavioral skill sets were desperately needed. By April, I received an itinerary for my first tour to Iceland. Robin had convinced OFS to fly Dave and me to the Vestmannaeyjar project site. Due to other demands on our schedule, I was to go first, and Dave would join us a week later.

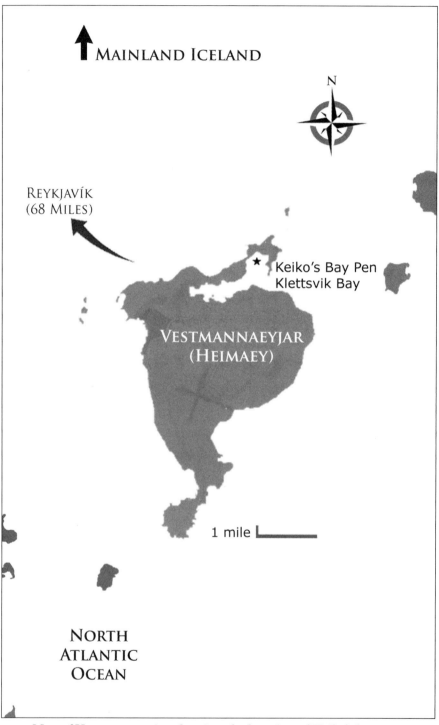

Map of Vestmannaeyjar showing the location of Keiko's bay pen.

Meeting Keiko and the Release Team

I once read that to truly experience Iceland, all one needed to do was to sit inside a walk-in freezer with coffee and a newspaper while burning a one hundred dollar bill. There is much more to Iceland than this spiteful commentary lends, though it is not entirely without justification.

Arriving into Iceland by commercial jet requires little effort to imagine what astronauts must witness on a lunar landing. The Keflavik International Airport is about an hour's drive southwest of the capital city of Reykjavik (ray-ka-vik). Flights originating from the United States arrive in the early morning, just as the sun crests the horizon. Coming in on final approach, there is nothing but volcanic rock as far as the eye can see. There are no trees, not many buildings to speak of and no color, not even the white of snow.

Keflavik is surrounded by pulverized black volcanic rock, slightly larger than gravel. Like many parts of Iceland, the area is frequently subjected to high winds, winds that single-handedly challenge all forms of vertical existence leaving behind a harsh and uninviting landscape. I would soon discover that the winds around Keflavik were only a mild introduction to the North Atlantic.

The Keiko Project was located in a small island chain southwest of the mainland referred to by foreigners as the Westman Islands or better known to locals as Vestmannaeyjar (vest-man-air). Keiko's base of operations was located on Heimaey (hi-may), the largest and only inhabited island in the fifteen- to eighteen-island chain. There are only two ways to get to Heimaey: by ferry or small plane. Both terminals are located near Reykjavik and require a very expensive one-hour cab ride from the international airport.

Weather permitting, the preference was to catch the twenty-five-minute flight to Heimaey onboard a nineteen-seat turboprop commuter plane. Otherwise the only remaining alternative was the four-hour ride to the small island aboard the Eimskip ferry.

Unfortunately, when the commuter flight wasn't flying it was usually due to severe weather conditions (a fairly common occurrence). This also meant the ferry crossing would be on par with Disney's "Mr. Toad's Wild Ride," with the rough seas tossing the ferry like a toy in a bathtub. As luck would have it on this maiden voyage, weather and schedules were on my side—I made the commuter flight to the island. Nevertheless, throughout the remainder of my involvement in the project I would get to know the ferry quite well.

I arrived in Heimaey at noon on April 27, 1999. My first dose of Icelandic adrenaline came on the small plane's turbulent landing. I am not a fearful flyer, but nothing about this approach was reminiscent of any landing I had experienced before, even in the smallest of planes. It just so happened that the runway, positioned in line with the prevailing winds, was not in line with the prevailing winds this particular day. The pilot had to approach the runway into the wind and almost perpendicular to the short landing strip, accentuated by sheer drops on both ends. At roughly 250 feet or so above the runway, the pilot spasmodically pitched the plane hard to port and dove toward the ground. Immediately, I was looking out my window and could see nothing but asphalt. I might have lost everything in my system, had there been anything in it. As fortune would have it, I don't eat when traveling, a complementary quirk that has served me well.

Eventually this signature landing style would become comfortable, and I would boast that Icelandic pilots must be the best pilots in the world. But for this once, it left my knees knocking and concluded an adrenaline-packed welcome to Vestmannaeyjar that was more apropos than I would yet realize.

Robin met me at the small island airport, and we embarked on a driving tour of Heimaey. The island was fascinating: in many ways seemingly inhospitable, but also beautifully quaint. For someone who (at the time) had not traveled the world, it was certainly an interesting place to cut my teeth. My first inclination was to believe that I had walked right onto the pages of a *National Geographic* pictorial, which after all, wasn't entirely unlikely. It had every element of the old world feel complete with "rugged ole" fishing vessels and "rugged ole" fishermen.

Much like Keflavik, Heimaey had no trees to speak of; there were the occasional saplings planted on the leeward side of individual homes and that was it. The houses and buildings were mostly white with colorful tin roofs of red, white or blue, huddled together as if to shield each other from the pounding winds that made Heimaey their playground. It took a mere glance on aerial approach to discern the island's most valued attribute, its protected harbor, from which the town of densely packed buildings and homes radiated outward. Houses appeared more like shelters disguised as homes. Their stalwart construction put the Three Little Pigs' handiwork to shame.

Icelanders that call Heimaey their home are proud of the island's many Viking charms. Among them it boasts the highest recorded winds of any populated area in Iceland. A remote weather station on the southernmost extent of the island had routinely, and I stress "routinely," documented sustained winds of more than 140 miles per hour; the most notable records cataloging speeds in excess of 200 miles per hour.

The heart of downtown consisted of a couple of cross streets lined with small shops, a geodesic-shaped grocery store and various multipurpose office buildings. Of course the majority of the town's economy was centered on fisheries. This was a niche community of nearly 5,000 Icelanders that was, by short description, a remote fishing village (or a drinking village with a fishing problem). Almost fifteen percent of Iceland's fishing exports came from this small town on Heimaey. It was a lifestyle and culture that would in ways provide both advantage and menace to the Keiko project down the road.

The sometimes extreme elements of the far North were not the only adversary that lent to the island's rich character. Vestmannaeyjar is the home of two notorious volcanoes, Helgafel and Eldfel. Both seem to tower over the small island town with an ageless indifference, yet no rational islander feels any indifference toward them. In 1973 Eldfel erupted and after blowing off steam for five months left the landscape forever changed. There is no shortage of locals in the Westman Islands who remember this event firsthand. Over the following many months I spent in Vestmannaeyjar, I would hear harrowing tales of the violent eruption. The lava that erupted from Eldfel covered parts of the town and increased the island's mass by nearly one fifth. Still, Vestmannaeyjar overflowed with an otherworldly charisma, not exactly what one would expect in a place called Iceland.

Part of the island created by the eruption in 1973 now serves as an overlook from the mouth of the island's harbor. At the end of the short tour, Robin took us to the overlook to see the operation from a more full view and to briefly spend time alone before meeting the staff. From here I could look directly north and see the bay pen, the base of operations and Keiko's temporary Icelandic home. In order to step out on the overlook, I had to open the car door for the first time since arriving, which was promptly ripped from my grasp by the winds funneling across the elevated observation point. *Stupendous,* I thought, not forty-five minutes on the island and I had already made my "mark" on the project. The door of that truck would never work smoothly again, popping and resisting each time it was opened or closed. *Nice,* I thought. *I'm sure that will gain favor.* Besides the wind, the weather was impressively mild, about twenty-eight degrees Fahrenheit; not the frigid biting cold anticipated this close to the Arctic Circle. The overlook appeared more like a flat spot amidst the volcanic rock than any tourist vista, remiss of any visitor amenities. We were the only patrons of its unearthly view that afternoon.

Waves crashing against the cliffs to the east of the overlook immediately commanded my attention. Despite the fact that the rise is well above sea level, there were geysers of saltwater being thrown high in the air above the edge of the drop-off; striking literally and figuratively. Opposite the overlook, the bay pen was nestled in a U-shaped bay just inside the mouth of a long, somewhat narrow channel. This was the entry channel to the well-protected harbor at the heart of the town.

The bay itself was surrounded on three sides by cliffs that shot angrily straight up from the water and towered hundreds of feet over the bay. My first thought was about noise. I had seen a multitude of large ships and boats moored in the harbor on our drive through town. It was surprising to me that the bay pen was situated so close to the shipping channel. Marine mammals have very sensitive hearing and can hear sounds across a much broader range of the frequency scale than human ears can appreciate. In marine mammal circles, it's often said that they "live in a world of sound." The position of the bay pen placed it right in what appeared like a giant parabolic echo chamber. *Surely the shipping traffic noise had to be detrimental to Keiko?* Overloaded with sensory input of the surroundings myself, I quickly became distracted with the bay pen, the bull's-eye of Klettsvik Bay.

The structure was quite simple and from my vantage point on the overlook, appeared incredibly small. It wasn't small, but relative to the cliffs, the massive bay and the mouth of the channel, it looked like an oddly appointed fish hatchery about the size of a tennis court. In actuality, the bay pen was almost the length of a football field and nearly seventy feet wide. The configuration was that of two octagonal circles joined by a smaller square pool in between. It did not narrow in the midsection; rather this is where the bulk of the deck space existed surrounding the small joining medical pool (or "med pool"). I could see two boxcar-looking structures placed opposite one another across the med pool. They looked as if they were balancing the bay pen from side to side. The pen's length was situated north to south in the bay, exposing the south circle to the shipping channel. I could not make out a killer whale in the pen. I shouldn't have expected to, the light was fading and we were roughly half a mile away.

Over the last week, Robin had been engaged in ongoing discussions regarding our involvement in the project. He was at an impasse. On the overlook, looking out at the bay pen and shouting through the wind and crashing waves, he shared with me the primary roadblocks to our proposal.

"They don't believe that behavior has anything to do with the reintroduction," Robin started.

I was stunned. "I'm not sure I understand. What do you mean that behavior has nothing to do with it?"

"It's the level of ignorance running the project; they don't have the tools to understand what to do next or how to prepare Keiko. To some degree, and what I don't know, they thought Keiko, once in Iceland, would show them the way. I think the only way to enlighten them is to slowly introduce them to what is needed for Keiko and explain the process in simple terms. But we're going to have to keep it simple . . . even Jeff doesn't consider behavior a part of it."

Jeff Foster, lead project manager, had a degree in psychology and a background in collecting wild killer whales, but he had never been involved in their training beyond that of initial acclimation. I didn't even know how to respond to Robin's comment.

"Then why am I here? I mean, why did they agree to fly me up here?"

"Because I insisted."

"Great, so I'm the black sheep that no one wants here or agrees with? Nice first impressions."

"We'll be fine."

"What's the plan?" I asked. Both of us had turned our backs to the wind, shoulders hunched up around our necks and hands shoved in our pants pockets. Robin's jacket hood had flipped up over his head. He peered around it as he continued.

"We'll meet with Jeff and Jen this afternoon and then get out to the bay pen tomorrow morning." Jen Schorr was Jeff's right-hand and the lead organizer for research data collection. "Charles Vinick is the Ocean Futures executive vice president and the Keiko Release Project's chief operations officer. He's arriving Tuesday, and we'll spend more time then or Wednesday going over the proposal and behavioral strategy."

I looked out over the bay. This was not exactly what I had envisioned.

"That's almost a week! What happens now? I mean we haven't come to an agreement or been hired, right?" The wind was modulating and had dropped at that instant. I was still shouting, and the sudden overcompensation sounded like an outburst. Given my sinking stomach, maybe in part, it was.

"No, but I'm pretty sure they want us onboard. The issue is going to be how they hire us. From my conversations with Charles on the phone, they don't want our company; they want us to work for Ocean Futures directly."

"Does anyone on-site know that we're here representing our company?"

"I think Jeff does and Peter might, but only because I know Jeff received the proposal. I don't know if Peter or anyone else knows. In fact, I don't think the rest of the staff has any idea why we're here."

After Dr. Cornell had specifically contacted Robin about the project and following his first tour and evaluation, we had sent a formal proposal on our company letterhead. The proposal created an awkward conundrum between us and OFS leads. Lanny had intended to hire a person, not a company. To them, it must have seemed we were trying to take over the project and their jobs.

I was wearing a denim long-sleeve shirt with our company logo over the left breast pocket. "What about this?" indicating my shirt, "should I cover it up? You think it'll confuse matters?"

"No, it's fine. Jeff knows why we're here, and I haven't tried to hide it with any of the staff."

We had noticed the ferry heading into the channel as we talked. It was now passing right between us and our line of sight to the bay pen. This was no small ferry; it seemed more of a full-on ship to me. Again I wondered about the noise. Not wanting to get off track, I checked the thought, convinced that I'd remember to ask about it later.

"I haven't really worked Keiko, just observed, and that's all I've really told the staff . . . that I'm here to observe the operation and help where I can," Robin said.

"So in the time you've been here and working Keiko, they haven't asked who you are or what you're doing here?" I pressed, completely dumbfounded.

As Robin explained, I learned that no one in upper management had communicated to those in Heimaey who we were or why we were brought to Iceland. For weeks we had been toiling with the makeup of our proposal, Dave and I in Florida, and Robin dissecting operations in Iceland. As a result, here I was now in Iceland and expecting my arrival and purpose were common knowledge. I didn't understand why the whole of it seemed so secretive. To me, it was simple. As the wave of realization came over me, I felt suddenly awkward. At best, I was an unknown and unwelcome visitor nosing in on their territory, and they had no idea why.

"So what do you want me to do until Tuesday?"

"You need to focus on Keiko. I want you to get a good read on him before we meet Charles. We'll go out to the bay pen with the opening crew tomorrow. We'll both spend the day on the pen and watch sessions. I've got some additional ideas to add to the proposal before Charles gets here. You're not going to believe it, it's pretty amazing, but the way they treat Keiko . . . it's like he's a big pet." The analogy was not the first of its kind I had heard, but used in this context to describe the release of a long-term captive whale, it was as chilling as the cutting wind.

So as not to be gone too long, Robin wanted to get back to the hostel, the living quarters for the frontline staff. Besides, we were both starting to force our words through clenched and chattering teeth. "Let's head back and put your stuff in the hostel, then I'll introduce you to Jeff and Jen."

The Hostel

The hostel was far from the bare-bones travel stops I had heard existed throughout Europe. This hostel was amazing. They had somehow leased a dormitory-like building from the local fire department. It had an entry foyer with a couch and a couple well-worn but welcoming chairs, a full kitchen, men's and women's bathrooms with multiple showers in each, several dorm-style bedrooms, and an enormous common area reminiscent of a small gymnasium with a pool table in the center. Near the far end and close to the kitchen was a dinner table fit for twelve with a white dry-erase board mounted on the wall right behind it.

There was a Partridge Family-meets-NASA feel about the place, warm in some ways, technical and clinical in others. Camera and recording equipment dominated another large table, and the entire front wall of the common area was covered by winter gear from parkas to fleece undergarments and even the occasional dry suit for diving in frigid waters. Most of the space against the wall was packed with Mustang survival suits—bright orange full-body survival suits that made the inhabitant look like the Pillsbury Doughboy no matter how thin the person wearing it. The suits were not much smaller hanging on the rack.

I don't know what I expected, but this was not it. In the small world of fieldwork, one does not naturally assume that a project is well funded or that everything needed to do the job is actually provided. On the contrary, minimum creature comforts, the sharing of gear and a constancy of fighting for equipment and funding are the norm. Here, this was not the case. I had never been involved in animal-related fieldwork as well-equipped as the Keiko Release Project.

Congregating in the front foyer, Robin introduced me to the first handful of rotational staff at the time. Jeff, in his mid-forties, was laid-back and right away disarming. In fact, he was similar to Robin in that capacity. Hair boyishly long enough to cover his ears, Jeff had a ruddy face that spent most of its time smiling, a smile that could easily turn into a shit-eating grin. He reminded me of that kid on the block that always finds trouble, the same one I couldn't resist hanging out with.

After getting to know Jeff over the course of the project, I would refer to him as a "whaleboy," the marine version of a Wild West cowboy. While effective in a multitude of ways, Jeff was a wrangler, the kind of person that shoots from the hip, nontechnical and nonanalytical but extremely competent just the same. He was gifted with an uncommon sense that allowed him to advance in his profession; yet, if asked to explain how he accomplished things he was often at a loss to adequately describe his actions or teach his skill to an heir apparent. In my experience, this was a common trait among those that do versus those that talk. Yep, "whaleboy" fit Jeff nicely.

Jen and her younger brother Greg were equally as disarming. Greg, the outdoor type, seemed adequately competent. His posture and enthusiasm revealed an eager-to-please youthfulness. He was young enough (early twenties) to be taken at face value with no ulterior motives or hidden agendas. A good-looking sort, Greg seemed to model the same boyish style as Jeff, sandy-brown hair over the ears but not quite below the neck. Greg's role in the project focused primarily on marine operations, making sure the bay pen was stable, piloting various watercraft and maintaining support equipment. I liked him right away. Jen and Greg were so well established in working together, it was almost impossible to tell they were brother and sister had it not been for the nights of heavier wine drinking when the childhood name-calling came out along with other gregarious banter.

Jen and I would eventually, simultaneously, become both adversaries and advocates. In as much as she was Jeff's right hand, I was Robin's, and we would soon be conspiring to keep operations between the two smooth. Jen looked to be about late twenties or thirty-two at most. Attractive and with shoulder-length light hair, she was dressed in jeans and a long-sleeve turtleneck that complimented her figure. Suitably thin, it always surprised me that Jen was not "granola" and did not allocate any of her time to working out. She was naturally fit, while her overall studious demeanor was juxtaposed by the occasional cigarette she would partake of while sipping coffee or wine, but never in front of Greg. Throughout my experience working with Jen, she was singularly focused on promoting and protecting the collection of data for any and all types of research that could be extracted from the project. I also found Jen to be genuinely concerned for Keiko's well-being.

There were others, but they had not yet returned from their duties on the bay pen. To my surprise and with no effort, I immediately felt comfortable with everyone I had met thus far. In retrospect, I suppose my expectation was to find something akin to the Berkeley radicals of the era. I had no reason to think this way; I had not heard anything negative about the staff on-site. In fact, I really hadn't heard much at all about them (certainly not from Robin, who didn't invest much time in character descriptions). My perception of the organizations leading the project had colored my expectations of those in the field. Once realizing that the people actually tending to Keiko were "animal-oriented people" I was able to let my guard down and felt more at home among professional peers.

The first night in the hostel, we all exchanged the usual small-talk introductions, drank red wine (a nightly practice on the project) and finally, turned in for the evening. The majority of our exchange had been fueled by my curiosity about the project and the people. For that night and many weeks yet to come I was in information gathering mode. But after a full day of travel and stimulation overload, sleep was a welcome reprieve. I had been on my feet for more than nineteen hours.

Keiko's Bay Pen

By five a.m. we were ready to go, clad in long johns, fleece outerwear and bright orange Mustang survival suits. The first stop after leaving the hostel would be the fish house, located in an old warehouse adjacent to the harbor. This is where Keiko's food was stored and prepared each day. In contrast to the enormous freezer warehouse in which it was located, the actual fish preparation room was not much larger than a walk-in closet. Every morning, the opening crew (typically two people) would bucket the fish that had been put in cool water the night before to thaw, weigh out Keiko's base (his total food allotment for the day), place it in steel buckets and cover it in ice for the trip out to the bay pen.

I was no stranger to "food prep" and immediately pitched in helping to scrub down the fish room and carry the four approximately thirty-five-pound fish buckets out to the truck. The two-story warehouse was a catacomb of freezers and was almost always deserted with little indication of human activity from one week to the next, although there was ample evidence of seemingly ghostly activity nonetheless. During daily ventures into the freezer building, we were often welcomed by creepy sheep heads, the decapitated remains of a healthy appetite for lamb in Iceland. Not far behind in ranking was puffin meat. Mounds of frozen and yet to be processed puffins would often greet us within the subzero structure. At night, when we would reverse the process of breaking out Keiko's food to thaw, the darkened warehouse full with carcasses proved to be the ideal setting for pranks.

Keiko's diet was identical to the whales' diet at SeaWorld. He was provided high-quality herring and capelin: 30 percent of the former and 70 percent of the latter, totaling approximately 120 pounds of fish per day. The only contrasting difference being that male killer whales I knew would typically eat between 200 and 280 pounds of fish per day. Killer whales require fewer calories in colder waters. Klettsvik Bay temperatures hovered around thirty-six degree Fahrenheit during winter months. Frigid water and Keiko's reduced activity level meant he didn't require near the bulk of food I was accustomed to feeding a whale of his size. Even so, we always took a little more than Keiko's set base amount out to the bay pen, in the event some of the food was dropped or lost in the wind, or if Keiko showed an unusually strong hunger drive.

With our survival suits pulled down to our waists and the arms tied-off behind our backs, we crammed into the back of the truck along with the fish buckets and a random collection of greasy marine gear and engine parts. Our next stop: the staff transport boat, *Sili* (sea-lee). It was a two-minute ride from the fish house, literally just a few hundred feet around the other side of the workingman's harbor.

The *Sili* was small, something you would expect to see on a calm lake in Florida, not the vessel of choice in transporting equipment and crew in the North Atlantic. She had an aluminum hull, about twelve to fourteen feet in length with a single outboard motor. Not much to write home about and overly crowded with even three occupants, but the *Sili* got the job done. On harsh weather days, the *Heppin* would take its place, a much stouter all-weather rescue boat designed expressly to thrive in Icelandic waters. On this particular morning, my first, all was calm and welcoming, and the *Sili* fulfilled her role without incident.

Rounding out of the harbor and into the channel, we were greeted by an ever-changing and inspiring scene. Jagged rock islands, just outside the mouth of the harbor, frame a distant glacier on the mainland. Defiantly emerging from the ocean's surface, the islands look tough, as if they are the last soldiers standing after a centuries-old battle with the elements. Like the bay, their walls are straight sheer cliffs that rise up well over 200 feet on all sides, making the island appear as an impenetrable fortress. Each one is topped with the characteristic Icelandic grass, tall enough to fall over in mounds, which from the distance appear more like an irregularly shaped surface covered in a thickening wet moss.

On the milder days, birds dominate the sky above Klettsvik and speckle the mossy grass, like salt sprinkled on green parchment. The sky was filled with birds of all types, sizes and shapes, and thousands of them, from the largest gulls I'd ever seen to the distinctive puffin and impressively large (albeit dull-looking) skua, a predatory seabird that commands its own air space wherever it patrols. All of this airborne activity gives Klettsvik Bay its own distinctive sound that, often paired with milder weather, quickly became a welcoming background ensemble.

The glacier, Eyjafjallajökull (don't even try to pronounce, unless able to vocalize on an inhale, this name like so many others in the Icelandic language is not within English vocal means), perched on the mainland over thirty miles in the distance, provided the backdrop and completed the most amazing commute to work I would ever enjoy. Every day the same islands framed the glacier, but somehow in the various lighting schemes experienced that far north, it always looked different. We never failed to be in awe of the glacier's beauty.

On my inaugural trip out to the bay pen, we approached the facility from the east. The norm was to approach from the west. In either case, it depended entirely on the wind. We always approached the leeward side of the pen whether that was east, west or north.

The wind owned Klettsvik Bay. Framed by sheer cliffs on three sides and so near the mouth of the channel, the bay acted as if a giant turbo scoop on the hood of a late model muscle car. If it was blowing at eighty knots offshore, Klettsvik Bay funneled the wind and amped that up to 120 knots sustained with gusts even higher. Not this day though. This day, my introduction to Klettsvik, it was deceivingly calm.

As we tied off to the bay pen, I could tell it was roughly the size of Shamu Stadium's main pool (a pool at SeaWorld Orlando of which I was vastly familiar, nearly 200 feet long and well over 100 feet wide), different shape, but about the same surface area for Keiko. My interest in the pen was short-lived. I would get plenty of time to analyze the structure later. Right then I was singularly focused on seeing Keiko. On the entire approach to the pen I had been scanning the surface for the familiar round black melon (or forehead) of a killer whale. The way the bay pen was constructed, much of the work area was two or more feet above the surface of the water and blocked the view. We reached the pen, tied up the boat and disembarked. At once I began walking across the middle bridge when finally I saw him.

The last two years of my life had been school and work. Purposefully intertwined, the two undertakings had parlayed into the start of a new business. Of great purpose and without pause, I had buried myself in the pursuit of altruistic possibility, tempered by equal amounts uncertainty. In this no-man's land, a by-product of the entrepreneurial endeavor, I had long felt a certain lack of security, as if my feet were not fully touching the ground. It was as if at times I couldn't get the right balance or traction. Amidst this feeling, enter Iceland and the Keiko Release Project: yet more uncertainty and now the supplement of vastly foreign surroundings. In the earliest hours of the morning the feeling is altogether reminiscent of childhood, when perhaps one stays too long following a sleepover at a friend's house. All is well, but in idle moments there is a longing for the security of familiar things.

This is where I was when I first met him. This was the feeling and state of my being that was so thoroughly vanquished at my first sight of Keiko. The all familiar black sheen; his movement so efficient and so smooth that barely a ripple escaped on the surface as he ascended; a recognizable breath that played like music to my ears; and then his gradual descent again leaving only the serpentine flow of his muscled back to follow. All this I knew. This I knew well. I immediately felt at home again.

Lesser Things

I couldn't stand it. I wanted to get my hands on him, see what he was made of... get a feel for his particular brand of bull killer whale character... look at his eyes. *What did we have to work with here?* I considered the variety of killer whales in my recent past. *Was he most like Kanduke, Kotar or Tilikum? Maybe Taku? Was he mischievous like Taima or a scary-smart Gudrun? Hopefully he wasn't a Winnie—that would never fly for a release.*

I had heard so much about Keiko's history, and studied every available morsel of information. But what cannot be read in a profile, history book or scientific paper is the kind of drive an animal has . . . whether or not the "lights are on." To be overly anthropomorphic, was he an extrovert or introvert? Outgoing or antisocial? Inquisitive or indifferent? All of this and so much more had a critical influence on a project of this nature, and I wanted to know all of it in one divine moment of enlightenment. Of course this wasn't possible, and my impatience would just have to be suffered. It would take time. Robin and the staff started me off with a tour of the bay pen. *Damn, can't you just leave me alone? I'm on the brink of a human to whale mind meld here!*

The pen was literally two donuts joined in the middle by a square rig that completed the necessary deck space for the research shack on one side and the dive equipment locker on the opposite. The "bones" of the bay pen and the outer rings were made of large, high density, foam-filled plastic tubes called HDPE, thirty inches in diameter. The black tubes were straight sections of various lengths and bolted together via enormous flanges on each end. These tubes provided the buoyancy and the structural integrity that kept the bay pen in one piece. However, flexible to a point, they also allowed movement. The bay pen would literally undulate and swell with each passing wake like a crowd at a football game doing the "wave," a contorting ripple that warped the pen from end to end and side to side. This happened in weather, of course, but also in calmer waters with the passing of shipping traffic in or out of the channel, often a dozen or more times per day.

The pen was anchored by a series of ship anchors, several tons each, that ran off in many directions. They were not visible from the surface, but later in the project, I would get a good look during maintenance dives. The first impression is that of a completely tangled mess, but they were, in fact, a systematically balanced and tensioned mathematical wonder that kept the pen from complete annihilation in pretty insane weather and currents. The operations team constantly worked to achieve the ideal tension equilibrium amassed between the maze of cables. If one side or line was off by just enough, the resulting imbalance could swiftly break the pen to pieces.

A very course grid work made of fiberglass called "Chemgrate" was laid horizontally about a foot or so above the structural tubes. This constituted the deck and made the bay pen walkable. The surface coating had to be super rough in order to provide a stable foot-grip, but if any ever fell on it, they might fare better dragging their face across a cheese grater. Outside of the main deck areas in the middle of the pen, the grate only provided about a two-foot-wide passage around the expanse of the two main pools. Exterior handrails made of the same high-density plastic kept us from being blown off the deck and into the bay, but nothing offered protection from taking a plunge into Keiko's side of the pen.

Underwater, nets hung from all sides beneath the structural tubes that formed the shape of Keiko's pools. They completed the pen's confinement perimeter. The bottom of the facility, about thirty feet deep, was also netting, but attached to the vertical net walls by a large concrete ring underneath, constituting the entire diameter of each of the two main pools. The rings weighted the net, maintaining the pool's shape and providing somewhat of a sea anchor to the structural integrity of the pen itself. At low tide, the suspended bottom of the bay pen was only a couple feet from touching the ocean floor. During high tide, it might extend as much as twelve to fifteen feet from the bay's floor, stretching the anchor cables to their fullest. There were times that this extensive variation occurred in minutes rather than hours at the hands of many violent storm surges that plagued Klettsvik Bay in the winter months. The befuddling matrix that formed the anchor system was actually the front line in the bay pen's survival.

The "research shack" was out of this world. Never in my wildest dreams did I expect to find such plush accommodations on the pen. Heck, I was even amazed that there was power, or at least hardwired power! I expected things to be run from a generator. Nothing doing, and although they had a backup generator the size of a Volkswagen behind the dive locker, the main power and a phone line were run along the bottom of the bay straight out from the town. They even had Internet access!

The light green research shack was like something found on a construction site, only this one was in much better shape. Measuring approximately thirty feet long by ten feet wide, there were only two doors to enter or exit: one on the end and one roughly in the middle of its length. There were windows, too, but only at the southern end providing views of Keiko's pool, the harbor to the southwest, and the cliffs to the east.

We entered through the door in the end and walked into a bunking room that doubled as a "wet room," an area for disrobing survival suits after the routine dousing from the elements. From there a small foyer and bathroom joined the wet room to the "dry room" where the staff spent a majority of its time. The dry room had a small kitchenette with sink, running water, a coffee maker, microwave and cabinets full of more dishes and kitchenware than I had in my first apartment. At the back end of the dry room was a bank of video screens, about nineteen in total, providing images from all around the pen, including a few from underwater. There was also audio recording equipment and a few hydrophone hookups that allowed the staff to listen to or record underwater sounds through submerged microphones.

One of the hydrophones was connected to a speaker, always providing the constant low-level underwater sounds, echo-like bumps and grinds of the ever moving bay pen. Under the window facing Keiko's pool, a low counter provided desk space and included shelves above. There was a lone computer for staff use on the pen sitting beneath the east-facing window looking out toward Keiko and the interior of the bay pen. There were even blue and white flowered Midwest-style curtains. Three cafeteria-like chairs completed the accommodations.

Very cool, I smiled. As a person from the animal field, I was not used to having all these work-related toys. After a brief tour of the bay pen housings, including an explanation of the records taken on Keiko, ethogram data recorded on Keiko's activities, and various other procedures and protocols, we went out on deck to watch a training session with Keiko, the "Big Man," as I would come to call him.

Thrashing

Stephen Claussen slapped the water's surface, the signal for calling Keiko over to where he stood at the pool's edge. Stephen was the lead trainer on this particular staff rotation. Stephen had gained his whale experience caring for Keiko in Oregon. He was full of nervous quirks. At times Stephen would unknowingly rub his hands together, one balled inside the other as if the evildoer in a cartoon escapade. Other times he would do it consciously, acting out the backdrop of a twisted comment. He was an immensely funny guy. His sense of humor was often a great and welcome equalizer in the middle of our newness, dampening the uncertainty pressed upon the staff. Stephen and I became fast friends.

The session was painful to watch. I had never seen such a slow whale. It was as if I was watching a fully loaded dump truck double-clutch through thirteen gears to get moving. Keiko, when he finally came over to Stephen, didn't even lift his eyes above the waterline. This posture is analogous to a person who "just-woke-up" dazed and with his or her eyes half shut. *Hello? Are you hearing me?* One can never be sure.

Stephen stood slightly hunched over, his chin almost on his chest as he peered down at Keiko. He nervously talked to him, his whistle bridge clenched between his teeth, narrating the more obvious while he pondered his next steps. (A "bridge" is an audible whistle signal that "bridges" the gap between the completion of a correct behavior and the whale receiving reinforcement.) Stephen's posture didn't lend much to a professional appearance. Instead, the way he carried himself made his clothes, the same apparel most of us wore, appear on him just a bit more disheveled.

Stephen moved ahead with his session plan, asking Keiko for a few behaviors. Among the menu of trials he gave the signal for a behavior they called an "innovative." Having no idea what I was watching, the session seemed to drag on with no end in sight. I was like a five-year-old sitting in church, fidgeting and struggling to pay attention. An eternity seemed to pass before the session was finally over. At last, Stephen stood upright, gave yet another signal with both hands, then promptly walked away from the session.

Afterwards I walked up on the bridge that connected the two main decks of the bay pen, the best vantage point to observe both the north and south pools. Keiko stayed right where they ended the session for a minute or so, and then turned slowly on his side violently throwing his head down, mouth open all the while and shaking one massive pectoral flipper wildly in the air. Rolling back to an upright position he took one explosive breath and started a slow counter clockwise swim in the north pool. W*hat the hell was that all about?* I wondered.

The odd behavior I had witnessed after the session ended was referred to as "thrashing" by the staff. To me, it appeared like frustration, a temper tantrum. Apparently, this was something that occurred with regularity and not necessarily during or immediately after training sessions. Sometimes it occurred seemingly at random, but always during normal daytime work hours. To my knowledge, the thrashing behavior never occurred at night or when trainers were ordinarily not present.

Standing atop the bridge I pondered every possible factor that might surround this odd behavior. I stared down at Keiko as he returned to his presession spot and slowly drifted to a stop at the surface. The sight of him broke my train of thought. If I didn't know better, I would have thought this was a pregnant female. I had seen fat dolphins before, but never a fat killer whale. From directly above, he looked like a giant guppy with his dorsal muscle ridge framed against a bulging undercarriage. Wow! I wanted to talk to Robin badly, but with the staff around and not wanting to insult my new acquaintances, there wasn't much I could say, at least not yet.

Ground Zero

Where I came from, killer whales were wickedly alert, fast and bright-eyed. In the training environment, a trainer gets what he or she pays for . . . meaning that behavior follows reinforcement. Further, dominant traits of an animal in a training scenario often betray the tendencies of the animal's trainer or trainers. They are clues to where the majority of effort has been focused (sometimes with purpose and sometimes completely unwittingly).

Think of a dog that jumps on its owner every day when the owner arrives home. The owner says "no" in a stern voice. Nevertheless, without fail, the dog continues jumping on its owner each day. Simply put, the owner is a reinforcing quality in the dog's life. Regardless of what the owner says when the dog jumps, the dog is getting attention from its owner, so the behavior is strengthened.

In this context, the word "reinforcement" refers to what happens immediately "after" a behavior to strengthen or increase the likelihood that the behavior will occur again. Not all consequences are reinforcing. Punishment for instance, has the opposite effect, reducing the behavior it follows. Specifically, reinforcement is said to "reinforce," strengthening the specific behavior it follows. If the consequence of a behavior causes that behavior to increase in frequency then that outcome is reinforcing (empirically defined). More to the point on Keiko, if "slowness" is reinforced, the result is a lethargic animal. On the other hand, if an atmosphere of high energy is cultivated by consistently reinforcing quick responses and attentive behavior, the result is a highly engaged and responsive animal.

"S^d" is the designation for Discriminative Stimulus. It is a specific signal that requests a specific response or behavior from the animal. S^d's can be hand signals, like sign language, or they can be audible signals or even environmental cues, like opening a door or turning on lights. Keiko did not know any audible or other forms of S^d. The vast majority of his learned repertoire was based on hand signals.

Behavior is a science; the application of behavioral modification (or training) is equally exacting. Among the many tenants of this practice, a trainer should not provide a signal or S^d asking for a response that he cannot reinforce or does not intend to reinforce. At the end of training sessions with Keiko, the staff would signal to Keiko that they were finished. It was the same signal that I had seen Stephen use. By giving Keiko a signal that they were "breaking" from the session (or ending the session) Keiko's caretakers presented an S-delta, a signal that indicates reinforcement is no longer available. In fact, theoretically, the signal itself becomes negative because it communicates to Keiko, "I'm leaving you now and taking my reinforcement with me." To Keiko's trainers it was a courteous, simple communication. To Keiko, it set the stage for frustration. Think of a toddler when he first recognizes the cues that Mom is leaving.

Schedules can also lend themselves to aggravation. If a session schedule is so routine that it becomes predictable, added to a "breaking" signal (delta), frustration can, and eventually will, escalate to its close neighbor, aggression.

There were many signs that Keiko's training stemmed from trainers with a background of pseudo-behavioral experience. These trainers, great of heart and talented in areas, never understood the science of behavioral modification, but rather, had techniques passed down to them through the school of hard knocks.

Another misguided construct involved that of the "innovative" behavior. This was a signal that was given to Keiko, and in response he could do (was supposed to do) whatever he wanted. The only requirement was that he could not do the same behavior twice in a row. Each time, he had to do something unique in order to receive reinforcement. They believed this was a fun interaction; that it stimulated Keiko to be creative and independent. In reality, it was yet another gray area of confusion for Keiko: no clear criteria, no clear direction and no consistency in the result. Spontaneity has its place in life, but not as a trained behavior. The two are mutually exclusive. The "innovative" was only the tip of the proverbial iceberg. In fact, most of Keiko's training interactions were, to borrow from Douglas Adams, "somewhat similar to but totally unlike" behavioral modification, leaving in their wake a host of aberrant and self-destructive habits in Keiko.

Years ago, as an up-and-coming trainer in the SeaWorld system, I was requested to appear at the office of the curator of Animal Training, Thad Lacinak. That particular day, Thad was perturbed by some stupid mistakes made by trainers which he had witnessed during a show; mistakes that he saw as creating confusion in the whales. A trend of confusion quickly leads to frustration, a predecessor of aggression. Unless corrected, this pattern of mistakes would ultimately get someone hurt. As one of the few "waterwork" trainers (trainers approved to be in the water with the whales), he challenged me with a simple question: "What happens if you bridge a behavior, the animal ignores the bridge, and you bridge again?" He was talking specifically about killer whales. At six foot three, Thad was an imposing boss. Adding intensity to his question, his normally pensive expression was painted with stern urgency while he awaited an answer.

The whistle bridge asks the whale to stop what it is doing and return to the trainer for reinforcement. Killer whales are top predators. Their hearing, eyesight and sonar abilities combine to make them the most aware animals I have ever been around. If a whistle bridge is blown, there's no such thing as "they didn't hear it." I responded to Thad's question, "That's a surefire way to get your ass chewed." I had passed the test. Continuing to provide direction when that direction is being purposely ignored is one of the fastest ways to ignite frustration and aggression in any animal (or person), especially a killer whale.

Yet again, this was another sloppy area in Keiko's interactions with his trainers. They would frequently blow their whistle, in effect, telling Keiko, "Great, that's it . . . now, come back and I've got a great reward for you." Often, when Keiko would not respond to the bridge, they would promptly bridge again, insisting that his hearing was bad. Highly unlikely; it was more probable that this "two or three bridge requirement" was the result of Keiko training (or ignoring) his trainers. Any other whale might well launch itself bodily out of the water, gaping mouth twisting to the side as if to grab the unsuspecting offender or at the least knocking them aside in a sweep of its head. Having witnessed this exact response to repeated bridging before, I can submit with confidence that once is enough to learn the lesson. I will never underestimate just how remarkably swift a large killer whale can be when driven to the point of frustration. If Keiko hadn't been so satiated with food, or been so numbed by this practice throughout his learning history, he might have reacted to these situations like a normal male killer whale and left the trainer with knocking knees for an hour or two.

These examples of conflicting signals, along with many others, were circumstances that should have totally pissed him off, yet the lack of aggression or even precursors to aggressive behavior from Keiko revealed another discomforting trait of this whale charged with surviving the wild: it was as if he had been "dumbed down" or dulled to the point of complete apathy.

In large part, the driving force behind Keiko's lethargy was not only poor training, it was compounded by diet. Because Dr. Cornell had mandated that a top priority for Keiko's release was to fatten him up, Keiko was completely satiated with food. Keiko simply didn't care whether food was offered or not. The only motivation to interact with his trainers was the stimulation they provided, and the break it offered him from an otherwise monotonous day, void of social contact with other whales, stimulating mental challenges or any other form of variability aside from changes in weather or current. Occasionally, when Keiko would not even care to come over, the staff would literally throw herring at him to make sure he got all his food for the day. Often, without moving, Keiko would just watch the herring sink to the bottom.

Imagine being so satiated after a Thanksgiving dinner that a nap is all that holds interest. Friends call and want to toss the Frisbee. Boredom begs for agreement, but motivation is stifled by an overbearing impulse to lie motionless, drifting in and out of consciousness. This was Keiko. This was the whale destined for release to a supremely unforgiving environment.

His fattened and lethargic condition had become a smoke screen that clouded any true evaluation of the animal. First, we would need to get Keiko on a workout regimen. We needed to get him moving and burning more calories, but not necessarily dropping his food intake in the process. Doing both at the same time might trigger the opposite result for which we were aiming. A drop in fuel at the same time we turned up his calorie burn could push his body to store more reserves, more fat. Robin and I couldn't contain our need to discuss the issues and eventually allowed ourselves to share our observations openly in front of and with the staff. During the following weeks and months, we would find ourselves constantly educating the staff on the basics of behavioral modification, normal killer whale behavior, nutritional dynamics and physiology. But talk is cheap; we would need to show them results.

Initially I spent a great deal of time with Stephen, who seemed willing to share the most detail regarding Keiko's recent past. We talked about the staff's affinity for getting in the water with Keiko and how that practice retarded progress toward rehabilitation for release. It seemed common sense to me that continuing playful in-water interactions with Keiko were not in alignment with the goal of release. Stephen offered that it was one of their only ways to keep Keiko stimulated. Still, I wondered what the world would think if they saw just how docile this animal was with his trainers in the water rubbing his back and belly. In a similar vein, we discussed Keiko's fixation with the blue Boomer Ball, which Stephen described in great comedic detail sparing no small amount of adult-rated analogy. I pressed him on the activities in Oregon, the rationale behind program directives, who had implemented behavioral protocols and details of what individual sessions were like.

According to Stephen's description, little emphasis had been placed on Keiko's learning. The primary stated goals of the Oregon phase had been simply to put weight on the whale and eliminate dependency on a slew of medications. In this capacity, they had certainly succeeded. Keiko was on nothing more than supplemental vitamins needed to replace the nutrients lost in the freezing and thawing of his food, of which he received a handsome quantity day-in and day-out. Judging by his enormous size, this amount was certainly more than he required, which had much to do with his lethargy.

Throughout the many hours I dwelled on the pen, I amassed more one-on-one time with the diverse staff. With each passing exposure, I learned about their past and how each had become involved with the project. The release team was divided into two rotations of personnel. Each team worked four weeks on-site followed by four weeks at home in the States. Every four weeks, a completely new staff would rotate in, occupy the hostel and take over the operation. Jeff led one rotation and Peter Noah the other. Although anyone on staff could and did work with Keiko at times, the primary individuals who attended to his needs, and whom I met on that first rotation, were Stephen, Karen McRea and Steve Sinelli.

I was shocked to learn that none of the three had ever worked with a killer whale before Keiko. Worse yet, none had any professional experience in animal behavioral sciences. For example, Stephen, known informally as the "director of comedy relief," had been a restaurant chef before joining the Keiko team. He proved to be a master of the galley, concocting some of the most exquisite Thai food I had ever consumed; this from the non-Thai assortment of raw ingredients available on the rock island. He, like so many providing Keiko's daily care, cherished his role in training and was all heart when beside the pool. Unfortunately his professional experience was in the kitchen.

When they conquered the moon landing, NASA was given almost unlimited powers to call upon and collect the world's most prominent scientists in rocket propulsion and lunar exploration. These innovators demonstrating a technological prowess and singular focus, eventually achieved the impossible. Here we had what consisted of an emotionally charged group of volunteers from the

Oregon Coast Aquarium leading the most ambitious animal release program ever conceived. If there was a silver lining to the lack of experience on the project, it was that they would be hungry for clear and focused direction. Or at least that's what was initially represented.

I was drawn to Keiko in many ways. He was a killer whale, one of the most amazing species of animal that I have ever worked around. Beyond that, Keiko himself was also strangely different from any other male killer whale I had ever known. There seemed to be no limit to his acceptance of anyone or anything. From all that I learned and was told, from everything that I saw, Keiko did not possess an aggressive bone in his body. His disposition with people was more akin to that of a big mellow Saint Bernard than any form of apex predator. These characteristics were intriguing and at the same time deeply troubling. At any other time or place, I would have been thrilled to be working with such a remarkable individual.

However, we weren't there to build a lasting relationship with the animal. We were there to prepare him for a life in the North Atlantic. That life would require diverse new skills. Of pivotal importance, it would require the absence of human relationships. No, in fact it would demand that he develop relationships with his wild counterparts, something over which all our preparation could have no influence. It was impossible not to like the Big Man, but beneath the excitement of meeting Keiko, ran a profoundly disturbing and dark current. I worried for this animal; I worried that what made him such an incredible animal to those around him would be his Achilles' heel, the downfall of his ability to succeed on his own.

The Enemy Within

E-mail: April 29, 1999

> *To: Alyssa*
>
> *Subj: Update (morning)*
>
> *We meet with Charles on next Tuesday or Wednesday. We are frantically getting our sh-t together. All is going well enough though. I feel so at home with a KW ... so very familiar ... it didn't take two days or even one ... I felt natural and at ease after the third minute. Great feeling (amidst the foreignness of everything else). This facility is awesome and the view is unbelievable ... beautiful weather today.*
>
> *Until then, I love and miss you. Mark*

Dave, my counterpart on the implementation of our outlined plan, was scheduled to arrive just a day before Charles Vinick, the chief operations officer for the Keiko Release Project. We would have only one day together on the bay pen with Keiko before meetings with Charles would begin. Although Robin and I had spent considerable time analyzing every aspect of Keiko's reintroduction and our proposal, I had much more hands-on time in the application of behavioral modification with Dave from our years of working side-by-side at SeaWorld. We each knew the other's strengths and weaknesses while also practiced in exercising each other's intellectual limits on the finer points of applied behavioral science.

From my own experiences, I had categorized three basic types of animal trainer: the "Relationship Trainer," the "Scientist," and the "Poet Philosopher." Relationship Trainers have a special knack for building relationships with animals. In other words, animals like to work with these trainers. More often than not, Relationship Trainers don't have a solid grasp of the science of behavioral modification and can't explain (very well) to others how they get results. Nonetheless, they are able to achieve an amazing rapport due in large part to their genuine affection and interaction with the animals in their care.

The Scientist Trainers approach animal behavior from a textbook perspective. They know the mechanics of how behavior is shaped. The Scientists don't have the best relationship with the animals and often face their own frustration; frustration that comes from the missing link provided through genuine affection. The best scientists recognize that developing a relationship with the animal they are training is a fundamental tool and requirement in the behavioral modification process, but this understanding is clinical. Cold application as a tool in shaping behavior, regardless of accuracy, does not a genuine bond make. Animals pick up on this, and their interest and motivation to work with the scientists is less than ideal.

Poet-Philosophers combine the talents of the Relationship Trainer and the Scientist Trainer. They possess a sincere desire to have a working bond with the animal, and they understand behavioral science. However, it's not enough to know the science of behavioral modification, there is an art form in applying these principles outside of the proverbial lab. Poet-Philosophers are great technicians, genuine partners with the animal and competent scientists. The Poet-Philosophers ultimately lead the field in animal training regardless of their title. They are often the silent and unassuming force that pushes the boundaries of achievement in animal behavior.

In a truly effective training environment, all three types of trainer are needed. The Poet-Philosopher leads with results and inspires an attentive animal, the Relationship Trainer provides a loving atmosphere that translates to relationship-challenged trainers and the Scientist Trainer maps the strategic path to the group's goal. The Scientist Trainer also plays an important role in teaching Relationship Trainers and young Poet-Philosophers the science of behavioral modification.

In many regards, Dave was heavily slanted toward the Scientist Trainer on this scale. Although relationship building with Keiko would be critical in the initial stages, this was neither a show nor presentation environment. Make no mistake, the importance of a solid relationship is not limited to a show or presentation environment. In our case it was just one piece of a multidimensional puzzle. At this early stage, the most vital aspect was mapping out a strategic plan with a vigilant eye toward the science of how learning occurs and how behavior is shaped. We couldn't leave any stone unturned. I was looking forward to Dave's input.

Excerpts from e-mail: May 2, 1999

> *To: Alyssa*
>
> *Subj: Updates from way North*
>
> *Being here puts things in a perspective that is difficult to describe.*
>
> *The facility sits right in the North Atlantic . . . you think harsh, and it is . . . they have sustained winds of up to 140 mph (it drug six 5-ton anchors about ten feet, but the bay pen held —and they then moved them back with two tug boats).*
>
> *The wind got up to 40 knots today . . . very interesting. It definitely starts things to rock'n. Dave and I worked Big Man (that's what I call him). I am already starting to fall for the guy. Anyway, Dave and I worked him in the wind storm . . . it was fun. Like doing a playtime in the heavy rain, only it was going sideways and it wasn't rain, it was the waves. It is definitely rugged up here, but this set-up is any animal person's dream. The only problem is they don't know the first thing about behavioral science or modification . . . and they openly admit it, often.*

> *The staff talks about him emotionally; no objective content. When Robin and I or Dave and I discuss behavior they start to glaze over. We have a looonnnggg way to go . . . you have to see it here soon though.*
>
> *We saw wild whales (KW's) less than a thousand yards from Keiko twice in the last three days.*
>
> *M*

Dave, Robin and I talked constantly. A tsunami of information and analytics were coursing through our minds each day. This relentless interaction never produced any abrupt revelations; rather, theories developed slowly as we reflected on the interaction of various applications, their intended results, the effects of unintended results and solutions for same . . . it was a necessary process that had no end.

The plan had to consider the incidental, accidental and indirect reinforcements that may occur in the process of shaping behavior. The reintroduction plan had to be highly sophisticated and involve myriad "what if" and "if-then" calculations. Beyond framing a comprehensive and plausible behavioral strategy for Keiko's release, we had the difficult task of explaining what that plan meant (translating it in layman's terms), and why it was at the heart of any chance of success for Keiko. At least at this early stage, understanding each step to be taken was a vital piece of the puzzle needed to foster the staff's dedication. Consistency in executing the plan would be paramount. How effectively we implemented each step depended entirely on the staff's grasp of each calculated measure as we slowly introduced change in Keiko's world.

Throughout the project, explaining the plan along with supporting theory behind it was an uphill battle. At that time, the managers in charge of the project did not believe that any reshaping or modification of Keiko's learning history was necessary. They believed that nature would take its course if they simply got him to Iceland. Yet after five months in Iceland, it was clear to even the most stubborn observer that Keiko was not making any notable progress. Other than watching almost every passing boat, Keiko was showing no overt signs of interest in the world outside of his bay pen. Thus far, and in the short term, this was the only aspect working in our favor.

Our interactions with all the existing staff were very positive up to that point. But then we were all reasonable and likeable people, them and us included. We were in the honeymoon stage of our tentative and new union; of course everything felt right. Nonetheless, our involvement in the project was not a given. Our challenge now was to determine how, formally, we would be involved. OFS had no intention of hiring a subcontractor beneath them; they were already a subcontractor of sorts to the FWKF themselves. On our side of the proverbial negotiating table, we were equally as stubborn and had no intention of splitting up our company to become employees of OFS. Perhaps it was some measure of pride, but it was also a commitment to the goals we had set as a new organization with what we believed to be a cutting-edge philosophy. The prospect of abandoning our company in exchange for employment within the rank-and-file radical organizations was almost too much to stomach. Charles Vinick arrived on Monday, and we met that evening over dinner.

E-mail: May 3, 1999

> *To: Alyssa*
>
> *Subj: Today…*
>
> *Winds of 100 mph and gusting around the pen at 110 mph. The staff goes about their business like nothing is happening. We watched a 200 ft. boat pitch in the waves like a toy. They say it is fairly common to see 40 ft. waves. I have not seen that yet; however, these were pretty awesome. It is truly a harsh environment. This weather pattern is supposed to last for another couple days.*
>
> *I am not very optimistic that we will be involved in this project. We met with the bigwig today. Went to dinner, drank some wine, he asked questions we answered them . . . typical GM type but very clever. He finally pushed the issue to hire us on as OFS*

> *employees and forego our company on the premise that this undertaking is bigger than all of us and together we had the "right stuff" . . . he also said he was interested in what we could accomplish together long-term . . . beyond Keiko. A tactic.*
>
> *Perhaps I am stupid to lose an opportunity like this one?!? We meet all day tomorrow starting at 0900 to go over the reintroduction and . . . I think he will try to sell us on the OFS employee thing again. We shall see . . . I guess the good news is at least he seems to recognize the need for our expertise.*
>
> *Hope all is well at home . . . I miss you ABS.*
>
> *Over and out,*
>
> *Mark*

I liked Charles from the moment I met him. A true gentleman in his early fifties, Charles' well-groomed salt-and-pepper beard and neck-length hair of similar spice fit his demeanor. Longer hair and a beard seemed to be the trademark style among the project's outdoorsy masculine leadership; yet Charles, departing from the seeming uniform persona, took on a more professorial air. His well-practiced, calm and collected stature produced a very businesslike impression. It seemed clear that this presentation was active and deliberate from the also apparent pride with which he carried himself. A smaller man, his dark complexion and sunken cheeks completed a very serious-minded appearance aptly fit for a board room. Nonetheless, Charles transitioned well into the fieldgarb typical on a project of this nature.

Charles was well spoken with an authoritative command of the English language. He was adept at clearly explaining himself and his objectives. This made negotiating much easier than it otherwise might have been. Despite the fact that he tenaciously held his own and refused to back down, I respected Charles for the way he treated our position at a difficult time. What could have been a very contentious exchange was maintained at a mutually respectful level. Charles never directly insulted our principles or our drive to protect our own company goals. If he had, the outcome might have been very different. In spite of these productive exchanges with Charles early on, there were other variables that muddied the waters and sidetracked negotiations.

The night of May 4, 1999, Robin and I were to meet with Charles a final time in hopes that we could come to an agreement. Early that afternoon, long before the scheduled dinner meeting, a Web posting made its way through the ranks of the FWKF board, eventually plaguing Charles with questions and doubts about bringing Robin and me onboard. A long time anticaptivity activist and vigorous proponent for the release of Lolita (a killer whale living at Miami Seaquarium), Howard Garrett had published the claim that SeaWorld trainers were in Iceland intent on sabotaging the release effort. In his posting he alleged that Robin and I were agents only interested in the project to secure Keiko for the zoological community. Failing in his first attempt, he went on to insinuate that any success would be credited to SeaWorld via our participation.

As outrageous and ill-advised as the assertion was, it nonetheless resulted in numerous phone calls from various members of the board fearing Garrett's on-line nonsense might be true. Robin had relayed the information while I was still on the bay pen that afternoon. Though I was not surprised by the elementary tactics Garrett used, the reactions of the board brought into question whether anyone involved actually cared about the animal near as much as they cared for their own credentials.

At dinner that evening the SeaWorld association and Howard Garrett's conspiracy theory certainly hindered the more pressing and immediate aspects of our discussion; however, being the level-headed man that he was, Charles did not let it delay progress for long. He knew the Howard Garrett claim was a cheap shot from the radical sidelines. After some relatively minor deliberation on the subject, he dismissed the notion and left it with, "I'll handle this with the board. Let's move on." Ironically, this was the beginning of what would become Charles' constant burden over the next sixteen months of our involvement.

I was hopelessly committed to making this work, even if I didn't know it then. The issue of becoming an employee of OFS plagued me more than it should have. The ease at which Garrett introduced turmoil among the more imaginative FWKF board members, could not be underestimated. It reminded me of the organizational challenges lurking in the shadows that only complicated the very real trials facing Keiko. I had now spent the better part of three days away from Keiko and neck-deep in consternation. Finally, my wife, who represents the better half of my judgment, spelled it out for me.

E-mail: May 5, 1999

> *Subj: more . . .*
>
> *From: Alyssa*
>
> *I know you are in a very difficult negotiation. As for my own advice, I offer the following input, not to pressure you either way, but rather to give you more information to consider.* **Don't be prideful.** *I fully understand not wanting to dismantle an organization, but don't be impatient either. This experience is right up your alley. No one is more qualified than you to do this; it is a natural progression and culmination of your work and school background. You can still have [your company] and all of its potential in the future. You do not need to be in such a hurry.*
>
> *I know how exciting this project is to you. It feels good, not only to work Keiko, but to work at the top of the game. You are MADE for this project; I'd hate to see you give up all of the positives in the name of something that may or may not be that important down the road.*
>
> *Understand the other chess player. You mentioned some very valid fears of Charles and OFS. Realize that these are solid trepidations that freaks like Howard Garrett are reinforcing. You need to get a line out to Ken and Kelly Balcomb, so that you don't have the peanut gallery advertising incorrect info. OFS is not to be looked at like an adversary.*

> *Finally, whatever you do, I love you very much. Be safe and warm and know that my heart and mind brim with pride at being your wife . . . your soul mate.*
>
> *Your Wing, Aly*

At times I believe that we men are in a constant state of self-destruction and that the only thing preventing us from sinking into oblivion is the right woman. My wingman or "wingwoman" had helped me turn a difficult corner. She was right; I wanted to be on this project and turning it down because they wouldn't hire our company would have left me with haunting regrets. This project wasn't going to last forever, and I could return to our company and its goals in the future.

That morning, Dave and I had gone to the bay pen to distance ourselves from the meetings and hopefully gain a fresh perspective. This is where I received the e-mail from Alyssa and after reading it, it was as if a demon had been exorcised from my psyche.

During that afternoon, Robin met with Charles as Dave and I were headed in from the bay pen. Famous for his skill at diplomacy, Robin was able to endear himself to Charles, and together they framed a sound understanding for our participation in the project. Among other agreements reached, Robin accepted the position as an OFS employee. He would be the new Keiko Release Project Manager alongside Jeff Foster. He also paved the way for my inclusion as the director of Animal Husbandry. This meant that I would be solely dedicated to Keiko and answer only to Robin and Jeff. Monumental change had to occur with this project and the way Keiko was being managed.

This structure allayed my final trepidations about becoming an OFS employee. That night I shared the outcome with Alyssa, to whom I owed my very involvement. The struggle and compromise behind us and with a singular focus now unbridled from lesser things, I was ready to put my full attention into every aspect of Keiko's rehabilitation.

E-mail: May 6, 1999

> To: Alyssa
>
> Subj: Hey
>
> *Came to an agreement with OFS today. I will be on a contract basis for 12 months as an OFS employee. The duration will be stipulated at a minimum of 12 months or as necessary until Keiko is released or relocated to a permanent housing facility.*
>
> *The next few days will be spent not only outlining the details of a behavioral plan but also creating my position description, title and responsibilities. Robin is the project manager. He answers only to the OFS board/executive committee. I will answer only to Robin . . . I also expressed to Charles Vinick my interest in business operations. He acted as if he was very interested in utilizing me in that capacity. Knowing Robin, and our past relationship, I feel good about being able to implement and influence organizational development. He said I was free to take on any and as much responsibility as I wished. We will see . . . there is tons of work to be done.*
>
> *Abs—I miss you terribly. I hope I can get used to this . . . it's tough. Foreign place, strange things . . . worst part is my best friend is away from me. Your words and philosophies are profoundly important to me, now and always.*
>
> *Love you,*
>
> *Mark*

Dave did not agree to employment with OFS. During the scope of the meetings with Charles, Dave had been hopelessly turned off. He wanted only to do what was best for Keiko, and to him that meant focusing on Keiko and only Keiko. The process of running an organization and a project of this size and scope invariably required discussion of rather unpleasant topics. After witnessing the "making of the sausage," Dave's impressions of the project, or more precisely the decision-matrix in charge of Keiko's release, left him with too many doubts about its eventual outcome.

Had we recognized the gravity of Howard Garrett's accusation, it's doubtful that either Robin or I would have continued on with our arrangement. I believe that Dave somehow felt this underlying threat to the project even if he couldn't put his finger on it at the time.

The fact was that Howard Garrett had not been the only one harboring a deep hatred and distrust toward us. Even within the board of directors, there were those that shared Garrett's sentiments. Far removed from the organizational turmoil over our involvement, many of our challenges in preparing Keiko for release were right in front of us, in broad daylight; measurable and factual. This is where we placed our focus.

We would not know until later, the real antagonist that threatened our ability to succeed was neither the harsh environment nor the behavioral obstacles that stood in Keiko's path; it was ignorance, dissention, and foul play behind-the-scenes working to erode everything we represented. It was the human element. Eventually we would discover the enemy within was not limited to the organizational headquarters in Berkley or Santa Barbara; even among us, there were those who monitored our every action, reporting to board members and poisoning the waters from day one of our involvement.

Regardless, for the time being, Robin and I had a clear path forward and felt unhindered by the office politics of the FWKF. In essence and in practice, we simply ignored the FWKF board, made possible by the project's geographical distance. From that day forth we approached the task of Keiko's rehabilitation with renewed energy, spending the next few days outlining specific changes that would be implemented immediately. By May 10, Robin had been in Iceland for more than a month. We decided he would head back home for some much needed R&R and to begin his rotation opposite me.

I would stay in Heimaey until sometime in June and get the ball rolling with Keiko and Phase One of the whale's rehabilitation.

Extracted and Condensed Summary from the original Release Outline as presented to OFS:

As presented to Ocean Futures Society, February 27, 1999
Excerpt:
A. Deprogramming
    1. Multiple-Baseline Design—extinction of unwanted conditioned history.
    2. ABAB/Reversal Design—Functional project relationship
B. Cognitive Restructuring
    1. Overt vs. Covert applied modifications.
C. Natural Environment
    1. Transference
    2. Modeling
    3. Response Contingencies.

ABAB: An experimental design in which behavior is measured during a baseline period (A), during a period when a treatment is introduced (B), during the reinstatement of the conditions that prevailed in the baseline period (A), and finally during a reintroduction of the treatment (B). It is commonly used in operant research to isolate cause-effect relationships.

A Legendary Place

The small island community of Heimaey has a calloused history that hardened its inhabitants resulting in a rich bond and brotherhood that suits the island-village well. Boasting a population of less than five thousand, everyone knows everyone, and strangers stick out like a redhead in a sea of blond, blue-eyed people.

    One can easily stumble upon stories of heroism and triumphant resilience against the elements when traipsing about town or frequenting the small tavern Lundinn on a busy night (which is most nights). Once beyond the initial cool gaze afforded strangers, little effort is required in becoming the favored guest of an overzealous fisherman more than willing to donate the Icelandic version of an Irish coffee while recounting a remarkable story with a matter-of-fact poise. Unless a glutton for punishment, I highly recommend avoiding such friendly gestures. A single dose of an Icelandic fisherman's Irish coffee, accompanied by a hearty shoulder grip is hardly an equitable exchange for a good yarn and may land the more gullible a thankless job onboard his fishing boat (a close encounter of which I have firsthand experience).

The town of Heimaey had a brand of small town charm uniquely its own. Townspeople were friendly, but something much deeper ran beneath the surface. It was as if there was a secret society behind the outwardly affable nature of most; though not apparent to a casual visitor. It took living and working in Heimaey to recognize the profound bond that the locals shared with one another.

Unlike any U.S. hometown claimed by the Keiko project team members, Heimaey had no crime to speak of. I doubted that anyone, a few I was sure of, ever locked the doors to his or her home. Children played throughout town with no parent watching over or worrying about their safety. On trips to the nearby grocery store, I remember being shocked that young mothers would leave their babies in strollers parked just outside the door while they shopped inside. I couldn't imagine the culture of security ingrained so deeply as to afford such comfort. But it wasn't the utopian safety or friendliness of Heimaey that gave its people a distinctive quality. It was something much deeper, more generational, something they were raised with and something individual to this island.

Iceland has perhaps one of the most pure cultures remaining. Very little outside influence on the society has taken place over the centuries. Its Nordic language, Icelandic, is a subgroup of Germanic languages and one of the oldest in the world. It is said that a modern Icelander could converse easily with a fellow countryman from the twelfth century, so little has changed about the language, the dialect, and the written word. The island of Heimaey is believed to have been first settled in 930 AD.

Dating back to the seventeenth century, Heimaey had been the target of the Turkish abductions, known as one of the most violent events in Iceland's history. Barbary pirates raided the island in 1627 capturing or killing more than fifty percent of the inhabitants.

Survivors were spirited off to Algiers and into a life of brutal slavery. These invasions happened again some years later when nearly 800 Icelanders were taken into slavery.

Between 1963 and 1973, several volcanic eruptions plagued the small town. In the most destructive of these eruptions, the volcano Eldfel began spewing volcanic ash and lava on the early morning of January 23, 1973. Almost the entire population of Heimaey had to be evacuated to the mainland within hours of the eruption's start. Many homes and farmsteads near the main fissure of the eruption were completely destroyed, either buried in lava flow or burned by flying lava bombs.

In addition to the violent history with Barbary pirates and challenges of the volcano fire mountain, the small fishing village has had no shortage of sacrifice at the hands of the unpredictable North Atlantic. Many families carry a scarred history of loved ones lost at sea. Local legends frequent the town's few drinking holes and lend to the otherworldly feel of the culture. In 1984, a local man swam over six hours in the frigid waters after his boat sank five kilometers east of Heimaey. No normal human could withstand the severely hypothermic waters for such an extended period. The man traipsed about the town and often greeted obvious outsiders who ventured into Lundinn with the same practiced taunt, "You a swimmer? You're no swimmer." A large man, his chides were often a drunken attempt to start physical conflict, but few patronized his invitation. Among the locals he's a living legend. There is also is a monument to Jon Vigfusson who scaled an impassable vertical cliff in 1928 to save the lives of his crewmates after their vessel stranded just offshore.

The island is surrounded by one of the richest fishing grounds in the region. Fishing operations that run out of Heimaey supply nearly fifteen percent of Iceland's total fish export. While stationing the Keiko project in Heimaey made it easy to feed a four-ton killer whale, it didn't, however, lend much in the way of moral support to the release effort. Most fishermen on Heimaey had no concept of what the Keiko Release Project was about. There are an estimated 5,000 killer whales in the waters surrounding

Iceland. To them, killer whales were a nuisance that interfered with fishing and competed for their resources. They often scoffed or laughed at us whenever we were fortunate enough to find ourselves in the same restaurant or bar the night a fishing vessel would return from weeks at sea. On many occasions the happenstance fisherman would take great joy in telling us that Keiko was no more than dog food and that we silly Americans were wasting money. The more serious and less social fisherman was not to be debated. At times, this contention in the community over what should be done with Keiko was palpable.

Not only was the Keiko Release Project considered a ridiculous venture by the hard-core fishing types, but also eventually the project would conflict with their livelihood and threaten the very food on their table. This time would come soon enough, but for now we had many more friends in the island chain of Vestmannaeyjar than enemies. Those friends welcomed every chance to teach us how to be festive in the long dark days of winter.

After warming up to their guests, locals are fond of sharing the three winter hobbies of Icelanders. They will deliver a robust slap on the back hard enough to throw one off balance and explain that during the twenty-hour days of winter darkness, their favorite pastimes are to "drink, fight and f-ck." Great shock value . . .only I'm confident the saying was not lacking of some truth, but rather born of much veracity. Such sayings, derived from fact, have a way of becoming legend in a Land of Fire and Ice.

In unlikely ways, the island village and the Keiko Release Project were perfectly matched; both somehow apart from the routine world surrounding Vestmannaeyjar. The island, the townspeople and their iconic guest comingled, forming a singular backdrop that defined an era unto its own in Heimaey's long and colorful history.

The uncertainty of our involvement now behind us, I found a deep connection with this new community, this new animal. After an intense first month, what seemed an eternity; it was finally time to get to work.

The Plan for Release—Phase I

If releasing Keiko was not as simple as moving him to Iceland, then what did a successful release look like? No one really knew for sure. Releasing an adult male killer whale like Keiko, who had been in the care of man for practically his entire life, had never before been attempted. However, the basic requirements for a cetacean to be eligible for release had been defined in 1993 by researchers for the U.S. Navy. These requirements had become the standard for evaluating release candidates. Represented here in their most simple form, the criteria were comprised of nine prerequisites that Keiko would have to meet prior to being considered viable for release to the wild:

1. Health: He had to be in good health. Keiko could not be reliant on any medication or veterinary assistance. His immune system would have to be able to deal with anything he might encounter without benefit of artificial support. Beyond the various pathogens or viruses that might exist in the wild, Keiko would also encounter bacteria and parasites found in live fish. His system would have to learn to cope with all of these new threats.

2. Physical Conditioning: He had to be in good shape (i.e., not injured or overly fat or thin) and able to travel great distances when necessary. Keiko had to be the equivalent of an ultra-marathoner or long-distance trailblazer.

3. Foraging Capability: Perhaps the most talked about in the general public forum, Keiko would have to eat live fish. It would not be enough to show us that he would or could eat live fish in a controlled setting. He would need to demonstrate

the ability to find and catch his own food. In killer whale societies, this is most often done through cooperative hunting. For this and many other reasons, finding Keiko a pod of wild whales that would accept him was paramount to his survival.

4. Normal Aversion to Man-made Equipment and Material: Keiko would need to be taught to stay away from man. That meant not approaching boats, docks, harbors or anything that constituted human activity in the ocean. He could not solicit for attention from humans during the inevitable encounters with man-made vessels at sea.

5. Avoidance of Humans and Human Contact: Keiko would need to forget his lifelong relationship with humans. He would have to replace his human relationships and prefer the kinship of his own kind. Conversely, attraction to human activity would be thereafter considered nuisance behavior and in regulatory fashion, mandate that intervention and permanent care be imposed.

6. Lack of Sensitivity to Monitoring Equipment: Basically, Keiko would be equipped with a tracking device, important for maintaining remote supervision of his well-being when he was successfully out on his own. All release plans require an intervention plan: an emergency plan to rescue the released animal should it get into trouble. Just because an animal may leave or be deemed "released" does not mean its release has been successful. Success would only come after surviving at least a season and showing further signs of thriving in its new environment or social group. Many times in the aftermath of release, even of rehabilitated wild animals, they become compromised after months or longer when back in the wild. Keiko would have to learn to accept a tracking device attached to his body.

7. No Behavioral Response to Acoustic or Visual Conditioned Stimuli: Keiko would have to forget everything he was ever taught by his trainers. It would take considerable time for forgetting to occur and likely only occur if Keiko had stimulation that replaced his training, namely interaction with his wild counterparts.

8. Normal Physical Capabilities: Keiko would need to demonstrate normal hearing, eyesight and sonar capabilities as well as normal immune response. He would have to show us the ability to dive deep and react to hunting or foraging opportunities. This

criterion required that Keiko had no permanent disabilities and that all his natural abilities were intact. It also meant that Keiko needed to know how to use them. For example, echolocation (locating non-visible objects by reflected soundwaves) was something that Keiko never truly needed in his life with man; he would now need to learn how to utilize his echolocation skills. Going back to Keiko's time spent in Oregon Coast Aquarium, there had been some question about his ability to echolocate. On several occasions Keiko had received deep gouges on his pectoral fins, flukes and his head (melon) and nose (rostrum). These were due to run-ins that occurred during the night with the underwater rockwork in the aquarium.

9. Social Experience with Conspecifics: This element was the capstone of them all. Keiko needed to be with conspecifics, otherwise referred to as "his kind." Beyond his health, normal physical capabilities and his acceptance of tracking equipment, many of the other requirements could theoretically be acquired from other whales through observational learning and pro-social behavior. Considering all the tools that could be employed in reducing Keiko's dependence on man, social integration with wild whales was the one criterion over which we had little or no influence. If Keiko was to be accepted by wild whales and also chose to be with those same whales, it would likely occur over time, after he and the wild whales developed a history together. In all our experience with animals of every kind, integration of individual animals into a new social group was tentative at best and often times treacherous. In short, it was extremely improbable that this mutual acceptance would be instantaneous, as was widely believed among the FWKF board and the millions of children who had emptied their piggy banks in response to the movie.

According to the staff, Dr. Cornell's masterminded release plan was a four-part venture (once Keiko was relocated to Iceland). The plan went something like this:

1. Physically acclimate Keiko to the seawater in Klettsvik Bay via the bay pen by exposure.
2. Stabilize Keiko's health (i.e., get him off any remaining special medications).
3. Fatten him up so that he has excess blubber to sustain him during the learning curve (whereby he spontaneously learns to hunt and forage on his own when hungry enough).

4. Literally fly Keiko out to sea in a sling suspended from a helicopter and drop him with wild whales.

The plan was heavily based on logistics with little beyond medical preparation relating to Keiko himself. Even the most inexperienced of the frontline staff knew this plan was a death sentence. They affectionately called this the "AMF Release Plan," short for "Adios, Mother F—ker." I don't know with certainty but suspect that the contention between Lanny's management of the project and the staff's unwillingness to do his bidding is what ultimately led to our involvement.

Our initial proposal to OFS provided only a summary outline of our suggested methodology but nonetheless departed from Lanny's simplistic cowboy approach in infinite ways. The plan was originally presented without detailed and up-to-date knowledge of Keiko or the capabilities of the field personnel. After only a brief time on the project, Robin, Dave and I had expounded on this initial outline adding immense complexity to its design. It did not take long to recognize the mountain that lay before us if Keiko was to be given any real chance at success. To even the most calloused, it was obvious that Keiko was as different from a wild killer whale as they come (or any killer whale for that matter).

At the most basic level, Keiko was a zoological animal. Throughout his life a dependency on man was created, in fact deliberately conditioned. Keiko's life with man was never about teaching him to be a top predator or a predator at all. His twenty years in a zoological environment were about building relationships with people. Keiko's facility in Mexico was substandard and affected his health negatively over time; however, that fact had little to do with his learning history.

Think of Mother Nature as the ultimate survival trainer. She doesn't give second chances often. She is seldom just and is ever-harsh in the immediacy of her verdict. When it comes to kill or be killed, there is no second place. In stark contrast, Keiko's life with man was a pampered one. His predatory skills were not necessary and therefore dulled to nonexistence. His trainers did not exact the same criteria as Mother Nature. If he had a bad day, he was loved or coddled, and excuses were made for his lazy response or lack of interest. He never had to navigate or find his food in order to survive. His life was never about what he *had* to do; it was always about what he was *willing* to do. He was fed enormous amounts and varieties of fish every day of his life through many varied human interactions, but little to no effort was required on his part.

Keiko never developed social skills with his own kind. His brief history with other killer whales often placed him at the bottom of the totem pole. In his early years at Niagara, he was frequently picked on and became a social outcast among other whales. After he moved to Mexico early in his adolescence, Keiko was never again with his own species. At times he shared his pools with other types of marine mammals, but he wouldn't see or hear another killer whale for a decade and a half.

Keiko's physiological development was also directly analogous to his behavioral upbringing. His immune system never had to stand on its own. If he experienced infection or illness, his system was bolstered by veterinary care and medication. Behaviorally and physically, he lived in a bubble. His bubble was not threatening. In fact it was designed to provide him with a carefree life—a life spent frolicking lovingly with humans. Yet Hollywood would have the world believe that this animal could instantly overcome this lifelong dependency, be whisked away on a flatbed trailer, plunked in the ocean and escorted to freedom where he would swim off into the sunset and magically rise to meet the challenges of a harsh and unforgiving world. No question, this notion was perhaps one of the most detrimental components in Keiko's actual release effort, and one that would relentlessly erode the project through to its conclusion.

We were not training a single behavior; we were conditioning a way of life, one that departed from everything Keiko had ever known. Keiko lived in the wild for no more than two years of his existence. In those two years, he relied almost entirely on his mother for food and direction. What memory of that part of his life still existed could not be measured, but whatever memory he might still possess was just as useless as if it had been completely erased. He had never learned to survive on his own, even before his 1978 capture. To be accurate, this was not a "reintroduction," but an "introduction."

Behavior is always in a state of change, always fluctuating, guided by the consequences that follow. At any given time, there are many and varied forces working against the intended goal. For example, people might learn to get off the couch and work out because they feel good and/or others make positive comments about their physique. However, if there is no direct recognition for their hard work, the relaxing state of staying on the couch is intrinsically reinforcing. This is a grossly oversimplified example. In reality there are dozens of factors influencing choices and actions.

For a new habit or behavior to take shape, encouraging reinforcement must initially follow that behavior at a high ratio. The process of learning carries with it a memory or reinforcement history, and the longer that history the greater the difficulty in reshaping behavior or replacing it with alternative or incompatible behavior(s). In the case of Keiko, we had two decades of learning history with which to contend. Consequently, the deck was stacked against our famous couch potato.

Neither the FWKF nor OFS had any experience or even a basic knowledge of zoological management. They did not have a grasp of the foundations that created Keiko, much less how to begin the process of reprogramming him toward a life of independence.

It is Change, Therefore We Must Fear It

At the start, our plan for approaching Keiko's reconditioning was laid out in a progression of several steps that would each be designed to meet a specific need or deficiency. Some needed to be implemented immediately; others would follow along with more material operational changes that would come only after Keiko had mastered the preceding steps. The first series of changes focused on both Keiko and the staff.

Step One: Eliminate all nonessential personnel and human activity around Keiko.

This step required the removal of as many ancillary associations with humans as possible. This important shift in his focus would need to start without delay and be progressively expanded through each stage of his rehabilitation.

This change was not well received by the staff or the FWKF. Prior to this suggested modification, everyone and anyone on staff worked with Keiko almost at will and regardless of their experience level or the presence of any program goals. This lack of focused direction translated into a very loose atmosphere. Keiko had become not much more than a team mascot. They failed to make the connection that this was supposed to be a killer whale, somewhat "similar" to the ones that hunt, kill and eat other mammals.

Visitors came and went from the bay pen at will. Staff members proudly escorted guests onto the pen to meet the famous whale. Even children were periodically brought onto the pen to play with Keiko. Though we struggled somewhat to reduce media access to the bay pen throughout the project, the "come as you will" policy was promptly ended.

As for the OFS staff or authorized access, we divided the teams into three different job areas; Behavior Team, Marine Operations and Research. It meant that we would identify a specific squad of people from each rotation, and only those individuals would work with Keiko hands-on in his conditioning. We kept the Behavior Team extremely small in order to minimize inconsistencies in individual training sessions. The first rotation would include Stephen Claussen, Karen McRea and Steve Sinelli. From the second rotation only Brian O'Neill and Tracy Karmuza would make the cut. Robin and I would split our time on-site straddling the two Behavior Teams and providing oversight, continuity and direction specific to Keiko. The rest of the staff was rolled into Marine Operations or research-specific duties.

Step Two: Eliminate extraneous trained behavior.

We immediately eliminated any trained behavior that did not provide physical stimulation or fulfill a specific husbandry need in the scope of the release program. For example, the innovative signal whereby Keiko was encouraged to do something creative was forevermore removed from the list of approved behaviors. It contributed nothing short of confusion in his conditioning progress.

Unfortunately, Keiko did not have a very extensive repertoire of trained behaviors (unlike whales at SeaWorld that learn and maintain more than 300 distinct behaviors). This mandate left scant few learned behaviors that could be used in exercising Keiko and keeping him mentally stimulated during the next phase of rehabilitation. More would have to be trained. New learning would also serve to awaken Keiko's sleeping mind, providing mental challenge in addition to physical work.

One highly unpopular amendment fell into this category that was far more difficult to implement than we ever anticipated: except for very specific medical or husbandry needs, no one would be allowed to get into the water with Keiko.

Upon this dictate, one would have thought we had committed a cardinal sin. Exactly why virtually everyone from the top levels down felt that long, in-water rubdowns and playtimes contributed to Keiko's progress toward independence was a mystery. What was clear is the huge contribution it provided for the *staff's* enrichment: FWKF board members, executive officers and others were allowed in the water with Keiko. It boggled my mind that this animal was slated for introduction to the wild and, yet, not one soul considered humans frolicking in the water with Keiko as conflicting with that goal. I guessed that it must have been common for researchers in the area to see wild whales swimming around the North Atlantic with Vikings on their backs? Even Charles, who evaluated most issues rationally and analytically, fought us on this change. As a result, it took much longer than it should have to finally eliminate in-water interactions from Keiko's regimen.

Step Three: Back to basics.

Establish and maintain consistency in Keiko's training environment and push his limits of higher energy behaviors and alert responses.

We would begin to concentrate sharply on Keiko's response to trained signals and push his threshold for active behavior in training sessions. In other words, we were going to make him a lean, mean, fast machine. But to get there, it meant ensuring that everyone who was left working with Keiko worked toward the same goals and communicated frequently. It meant constant supervision.

Step Four: Seize the moment!

We would immediately begin placing an emphasis on what influenced Keiko's behavior outside of training sessions. Getting Keiko in shape, both physically and mentally, started with cleaning up his training program and sharpening the basics. But training sessions only constituted about four hours of a twenty-four-hour day. What influenced Keiko's behavior the other twenty hours was of paramount importance, and it would become increasingly so throughout the next several stages of the release effort.

Without delay we instituted a constant "watch." This meant that we needed to observe Keiko and his environment at all times.

We had to consider every conceivable variable from boats to birds. Each situation represented an opportunity to either reinforce (or strengthen) behavior from Keiko that moved him in the direction of acting like a wild whale or to discourage behaviors or objects that we did not want to accidentally or indirectly promote. Humans were the sun in his sky. We had to be supremely aware of what we shone upon.

Accidental reinforcement can occur in numerous ways, such as starting a training session while Keiko is sitting still and inactive. That would be bad (starting something that is positive as a result of inactive behavior thus rewarding inactivity). Or say, if a boat were to approach the bay pen right after Keiko stopped swimming. The interesting diversion of a boat full of humans arriving at just the wrong time could reinforce Keiko for the act of being sedentary. The most common and seemingly innocent events could go completely unnoticed and yet have great influence on the ways in which Keiko's behavior developed. Clearly, we needed to start watching and actively managing Keiko's world 24/7.

Step Five: Get rid of the pacifier.

Last in the first series of immediate changes, we had to take Keiko's "pacifier" away. Keiko had a very large blue plastic Boomer Ball that stayed in the water in the bay pen at all times. When no one was interacting with Keiko, he sat motionless, like a log floating on the surface (logging) next to his three-foot in diameter ball. There were two rather pivotal reasons why this attachment was not helpful.

First, there are not many blue balls floating around in the wide open ocean, and it was doubtful he could take this one with him. Secondly, it was incentive for Keiko to remain inactive, floating all day next to his inanimate companion. The latter of these two was less obvious to the staff, who considered the artificial object a form of enrichment.

But there was more. The blue ball also provided Keiko with sexual stimulation. That's right, he became "randy" with the favored toy. The deviant affection was an impediment on too many levels to list. For the reasons stated and more, his attachment to the blue Boomer Ball could not be allowed to continue.

However, it can be dangerous to abruptly remove something that carries such a strong attachment. Therefore we began the lengthy process of slowly removing the blue ball from Keiko's environment, systematically rewarding appropriate behavior in its absence.

A by-product of this first round of adjustments was the onset of one highly toxic allergen detested by the human animal: change. As Stephen Claussen frequently reiterated: "It is change; therefore we must fear it!" As humorous as the statement was, the sentiment prevailed among the staff in the first several months of our management of the release effort. Slowly, as the stark reality of the new edicts were implemented, morale began to break down within the existing staff. The honeymoon was definitely over.

In complete contrast to the growing friction within the ranks of the reintroduction team, Keiko himself was actually beginning to respond, gradually at first, but eventually in leaps and bounds.

Two Steps Forward, One Step Back

Amidst the burst of initial changes we needed to increase Keiko's activity level, exercise him, and get him out of the funk in which he was mired. But in order to do this, we had to reteach him everything he knew. Start over. Why? Keiko's past training was exemplified by laziness: lazy responses to signals, lazy movements, lazy everything, and his trainers not only allowed it, they in fact trained it. In his past, this apathy had been a hallmark of his conditioning. He was the slowest and fattest killer whale Robin or I had ever seen.

In nature, a lackadaisical approach to catching food for the day will not fill the belly. We wanted crisp, awe-inspiring predator-like reflex reactions and alertness. In order to create an attentive, sharp, and quick animal, we had to go back and start from the beginning.

It almost seems common sense that an animal being prepared for a life in the wild has nothing to gain from continued relationships with humans. Those relationships would ultimately conflict with Keiko's long-term survival in the wild. But in order to move toward a goal of less human contact, we had to start with more human contact. The only way to physically jump-start this whale was to begin intensive exercise, and we had to do this using the trained behaviors that Keiko already knew well. More to the point, we were going to push Keiko beyond any work level he had ever known in the care of man. To be able to push Keiko out of this daze, I would have to be prepared to offer him something worthy of motivating him. I would need to establish a solid relationship with the Big Man that could withstand the test of the grueling months to come.

Molecular Movement

Stepping up to the side of the med pool, I glanced to my left and waited. I wanted to make sure Keiko was moving in my direction before calling him over. It was an unusually nice day, sunny and with very little wind. The air was still enough that snow accumulated on every horizontal pocket in the cliffs and nearly blanketed the nearby town. In the easiness of the afternoon the birds overhead were loud, a ceaseless cacophony of chirps and whistles.

Recognizing my best opportunity, while Keiko was moving, I bent down and slapped the surface of the water. Keiko didn't show any response to my slap. He casually continued his movement in my direction and eventually arrived right in front of me. He didn't lift his head or make any motion to acknowledge that I was there; rather, he only came to a slow stop with his nose pressed lightly against the side of the med pool structure.

I waited. Keiko didn't move. I could see he was watching me, rolling his eyes just so. But that was it. After a few moments I stepped back and walked away from the side of the pool. Behind me, Karen and Stephen had been watching.

"What was that?" Stephen asked with a smirk, finding the whole situation slightly humorous. After all that we had talked about during the last week, they probably anticipated that I would wield some magical power over Keiko and were expecting to be impressed. It couldn't have been less impressive. I had to jump up to touch bottom.

"I'm not accepting that—if he wants my attention he's going to have to sit up and give me his." I didn't mean for it to sound arrogant, but to Stephen and Karen it did. We stepped out of Keiko's line of sight and continued debriefing the short session.

"If we're going to get this boy moving we have to start looking for more energy . . . tough love," I said. Karen didn't like it.

"He came over the first time, though." It was more a statement than a question (and a measure of what they were used to). She asked what was next, more to test me than out of genuine interest. We were leaning against the north side of the green research shack, staying out of Keiko's view. Peering around the corner, I could see that he remained in the same position where I had left him.

"I'm going to wait until he moves away and is actively swimming, then I'll call him over again and see if he gives me anything worth reinforcing." I avoided a direct answer, hoping instead to illustrate my point. We weren't even scratching the surface, and yet I could tell Karen was already uncomfortable with how I was pushing Keiko. A few minutes later, Keiko was swimming in the north pool again. I stepped back up and called him over, again waiting until he was facing in my direction before giving any signal.

He approached much the same way and stopped, nose pressed against the side of the pool. I made no reaction and waited.

Behind me Stephen asked, "What are you looking for?"

"Any sign of movement—even at the molecular level at this point," I replied.

Finally Keiko lifted his head, and for the first time that morning his eyes appeared above the surface. I immediately reacted, giving the whistle bridge a short burst and tossing him a single herring. I moved quickly down to my right about thirty feet and slapped again. This time he came over slightly faster, not much, but arrived with his head up and looking at me intently. I made a huge difference in my energy and posture, reacting to Keiko's improved attention. I reinforced him again, this time with a small handful of herring, then broke away from the session.

The rest of the afternoon I repeated the same sequence of interaction, asking for a little more each time. I kept my sessions with Keiko short, each four to five minutes at most, and put a lot of emphasis on a variety of differing reinforcements or rewards. He never knew when I would show up or how long I'd stay, but the result was always intriguing. Sometimes I gave him a quick rub down on his pectoral fins using a brush. Other times I used the water hose to massage his flukes or spray on his tongue. I made quick and novel changes matching the level of his effort. That day was all about how far and fast I could push him. I wanted to find his limits, see how far he would go for the new guy.

Over the first week or more I made few attempts to work with Keiko myself, instead working him through the existing staff. There was sound theory behind this approach. We needed to teach the staff solid foundations of behavioral modification, and the best way to do that was to make them actually do the training, with guidance of course. I was dead wrong.

Largely because of the gray areas that came with sloppy and inconsistent interactions, Keiko's day-to-day life was rife with confusion. This did not lend itself to strong relationships with his caregivers. Also, most of the staff tried very hard to implement the instruction they were being given, but ultimately Keiko was learning faster than they were. I could no longer afford the delayed responses and second tries that came from working through the existing staff. More and more, I started working with Keiko directly.

It didn't take long to get a solid grasp on my relationship with Keiko. In at least one way, it was easy for him, as I was very clear about what I expected. If he met the requirement, great things happened. If not, nothing happened. It was black and white. Keiko responded to this clarity and consistency and began to excel. He was a quick study and began discriminating when it was me working with him. He knew from the start of each session that I would ask a lot of him, but he also knew I would meet his efforts with like energy.

The most effective trainers put a lot of planning into their reinforcements ahead of time and are ready to respond when the right behavior happens. Contrasting this elusive principle, it seems human nature to focus on what the animal is doing wrong rather than what's right. In the world I came from, the better an animal was doing, the harder the trainer worked in creating reinforcement. The best training sessions leave the trainer exhausted and dripping with sweat, no matter the frigid water or blustery chilled winds. There had never been much energy or planning put into providing exciting and diverse rewards in Keiko's daily training interactions. This aspect alone allowed me to stand out from the rest of the crowd. Fortunately, the staff was learning also, and the results we were seeing with Keiko quickly became motivating to everyone.

"Dancing Queen"

Animals in a training environment tend to "mirror" their trainer's energy level. Too many times I've witnessed a trainer plop down on his butt in front of an animal at the start of a training session only to have that animal take one look at Mr. or Ms. Boredom and promptly leave the scene for something more interesting, like watching paint dry. A tried and true method I had utilized in the past proved the easiest way to get trainers "off their butts" and infuse energy into their interactions with Keiko: music.

As tribute to one of our favorite Icelandic coworkers, Mr. Iceland, we blasted "Dancing Queen" by Abba across the bay pen during some exercise sessions (only in the early stages and weather permitting). The music was for the trainers, and their spike in energy produced night-and-day results in Keiko. Take a moment to picture the scene: "Dancing Queen" playing in the mind's ear, trainers flamboyantly engaging their entire body presenting to Keiko what would otherwise be a mere "hand signal" moving in tune with " . . . young and sweet . . . having—the—time—of—your—life. . . ." Music provided much levity but also produced some of Keiko's most effective "workouts." It was a pure, uncomplicated transition for the training staff to grasp and illustrated the fact that food alone was not a key motivator. By merely energizing the trainer's posture and creativity to react to Keiko's successes, we could extract some pretty spectacular energy from Keiko in return.

Bridging the Gap

In the beginning my time was dominated with sharpening the effectiveness of each and every tool in the training environment and eliminating "superstitious" beliefs about how learning occurs.

In every setting in which I had previously conditioned behavior with animals, we always "debriefed" after training sessions, discussing the various observations and identifying areas for improvement. Not only was this an important part of any environment where success relied on consistency, but it was especially important when the subject of that discussion was a killer whale. Small mistakes can lead to menacing consequences with the ocean's top predator. In this case, we couldn't let small mistakes undermine Keiko's needs.

I recall a conversation I had many times in just such debriefings, this time with Steve Sinelli and Karen McRea. We were sitting in the bay pen research shack having just finished a training session with Keiko.

Steve reminded me of one of Santa's elves. He was not categorically short, but close. Balding and with a close cut and well-groomed beard, his black hair gave no signs of graying and lent to his elfin appearance. Steve had a youthful energy that contrasted slightly with his age. More often than not, he wore loose fitting water-resistant nylon and fleece lined pants and a black fleece vest over his white turtleneck shirt. It was Steve's black and white uniform attire. He was confident, and like me, could be argumentative.

Steve sat in front of the computer but turned away from the screen, instead facing Karen and me. He had just finished writing up the session. Steve had worked with Keiko in that particular training session, rehearsing a behavior called a "fluke presentation." When given the signal or S^d for the behavior, Keiko would turn on his back at the surface and present his tail flukes to Steve who sat on a water-level floating platform. In this ventral position, blood samples could be easily drawn from the larger veins that run through the flukes. Typically, Keiko would remain in this position for three to four minutes, and sometimes as long as ten minutes depending on what was required. In past research on his breath-holding capacity, Keiko had held this position (and his breathing) for over thirteen minutes. This and many other husbandry behaviors were a normal and important part of Keiko's life. He knew them and performed them well. I described what I had seen during Steve's session and suggested a few adjustments.

"Tell me why you were using your whistle bridge during the fluke presentation behavior?" I asked.

I had learned to sit during these types of discussions. At six-foot-two it was easy to inadvertently bully a shorter person, and I didn't want to start off on the wrong foot. No matter, Steve was one of the older team members and was pretty comfortable in his own skin. Karen, not so much.

"We use a short whistle to let him know he's doing good and then a longer whistle to end the behavior when he's completed it." Steve replied instructionally.

I'd seen this *superstitious* use of the whistle bridge before and debunked it just as many times. It was a critically important foundation in establishing how behavior is conditioned. As importantly, everything in Keiko's world had to be refined, even the most seemingly innocuous training habits.

Karen didn't say much, but I could tell she was listening intently. With a degree in psychology, she knew the language of behavior but had never used it in an applied setting. Nonetheless, I found it more productive to speak in layman's terms, avoiding any confusion. "Okay, think of the whistle or bridge as a 'secondary' reinforcement. When Keiko first heard a whistle bridge it didn't mean anything to him. But over time, someone taught him that the whistle meant 'good' by following it *consistently* and *immediately* with various rewards."

They accepted this, but kept a poker face as if to say, *Duh, tell me something I didn't already know*. This much was cliché and often part of educational spiels at various training facilities. I was being ultra-elementary on purpose; I didn't know at what point along the line we would cross into new territory.

I continued, "The whistle *bridges* (I emphasized the word) the time between when Keiko completes the behavior correctly and receives his reward from you." There was more. "It also takes a snapshot picture in his mind at the *precise* moment that he has committed the correct response." I needed to drive the latter point home or they wouldn't get it.

"Imagine that Keiko is learning to do a jump in a specific place in the pool that you choose. How do you teach him to jump in that precise spot every time?" I wanted them to engage in the discussion.

Steve replied with ease. He had seen this and was not fooled. "You use a target pole and slap in the position where you want him to exit the water."

"Yes, but when you are fading that target, teaching him to 'remember' the spot without the help of the target, when do you *bridge* the correct response?"

Steve tested the water, "When he comes up in that spot?"

It was as much a question as a response. Steve knew I had a trick answer up my sleeve. He smiled . . . we were having fun with the discussion. Karen sat in the chair under the west window, happy that she was only indirectly involved. In my peripheral vision I could see that she was processing the question with a pensive look on her face.

"Not exactly . . . you want to use the bridge at the precise moment that he turns up from the bottom toward that spot. That's the moment you want to grab his *thought process* and say YES! That's it —you've got it!" I found myself standing to emphasize the importance of this precision tool, using my hands as if one were Keiko sweeping up to jump, and the other was the surface of the water. "That's when you need the whistle to be sharp and powerful—grabbing his attention."

But Steve was unsure of where this was going. "Okay, but what's that got to do with bridging his fluke presentation while he's doing it correctly?" he asked.

"Everything!" (I loved this stuff.) Continuing, I explained, "When you use the bridge and *do not* follow it immediately with a reinforcement or change, a consequence, some form of positive consequence—you are dulling a precision instrument—the whistle bridge. You are in fact desensitizing that bridge, reducing its value. After a while it is no longer a precision instrument but a blunt tool that has lost effectiveness. It no longer means anything when you most need it to."

Steve was not convinced. "That's why we use a long whistle during the behavior and a short whistle as the precision part."

"The long whistle doesn't mean anything to Keiko. It has no consequence, no change that gives it value. In effect, through generalization, you are only draining the value of the bridge as a whole—whether that's a long or short whistle or a catchy melody."

"Don't you think he knows it though, like we're saying *good boy—keep going?*" he asked.

"No I don't. By allowing him to continue the behavior you are accomplishing the same thing. You don't need the midterm whistle. Alternatively, you can rub his flukes, providing reinforcement while he's holding the behavior, and achieve the intended result. But you need to think of that bridge as a vital learning tool, and protect its value by making sure that it has consistently positive and immediate consequences. The whistle bridge has a very specific application; it's not a tool to be thrown around lightly and for convenience."

We would have many and varied conversations of a similar nature, discussing everything from the whistle bridge to transferring learning and reinforcement history from one environment to another and beyond. Although the prospect of teaching behavioral modification was thoroughly enjoyable, added to the vast needs demanded by Keiko and the long road ahead, it was exhausting. We needed "top gun" trainers that knew this stuff intuitively.

Do-si-do

The first rotation team and I were just beginning to find our groove when it was time to alternate the entire team. The second rotation would be sashaying in and taking up residence in the hostel, the bay pen, and taking the reins on Keiko's daily needs. The only holdover between the two rotations . . . me. This left no other option; I would have to pull out the most effective organizational secret weapon ever conceived by man—the "staff meeting" (in case there's any doubt, that was indeed heavy sarcasm).

As chance would have it, Jeff's alter ego, Peter Noah, was more of an organizational freak than I. Peter held an informal group meeting his second day, setting the record for 100 percent more meetings than Jeff had tallied in two months. And just like that, it was a completely different atmosphere in our quaint but peculiar hostel.

E-mail: May 9, 1999

> To: Alyssa
>
> Subj: Good Morning My Sweet
>
> New staff arriving and existing staff are on their way home. Quite an interesting exchange of issues. Lots of change happening . . . met the last supervisor level yesterday. I will be working with him through June. The weather finally laid down . . . sun is out today and the winds have dropped to about 19 mph. We have had this wind storm nonstop for the last several days straight. It is nice to finally have a little calm weather, not to mention seeing the sun. Speaking of which, sunrise is at 4:27 AM and sunset is at 10:03 PM, but it is never really dark. Between sunset and sunrise it just sorta stays twilight. Weird . . . I will take some more pictures today and try to send them your way by tonight. We had our first (though informal) staff meeting yesterday. They have not even had staff meetings on any regular basis. Given the nature (of the project) and safety issues involved with this operation it blows my mind that regular meetings and protocols are not in place. Soooooo much to do. I will work out my return date with Robin today and let you know my schedule.
>
> Until then, love is in the air,
>
> Marjke (Icelandic for "Mark")

Peter and I hit it off immediately. We spent his first full day back in Heimaey sequestered on the bay pen—just the two of us. Peter was now the acting on-site project manager. He wanted to know who this guy was running things with Keiko. It just so happened that Peter was a very analytical, left-brain thinker. In my past, I had been accused of being a sneaky-deep-down "Vulcan," favoring logic. We both enjoyed a good clean whiteboard.

Throughout the day we discussed Keiko, the behavioral science behind our proposal and how it could be effectively implemented against the many logistical and weather challenges in Vestmannaeyjar. The conversation covered every identifiable hurdle, including the limitations of the Behavior Team, some of the existing gaps in management structure and the oddball staff rotations.

At that time, I would guess Peter to have been in his early forties. Taller than average, he carried a large frame and were it not for his somewhat academic nature, would have been an imposing figure. As it were, Peter was an easygoing guy. He liked to talk things through ad *nauseam*. The staff liked Peter, but tired of his systematic management style and preferred Jeff if given the choice. It didn't take but the first forty-eight hours with Peter and the incoming rotation to figure out that this new group was the motley crew, the leftovers after Jeff had handpicked his preferred A-team.

A massive overhaul to the staff rotation and organizational structure became a priority. Peter had already recognized that a complete shifting of personnel without any overlap was detrimental to maintaining any consistency with Keiko, or any other operational detail for that matter. We conspired to imagine what could be done and what changes would need to eventually take place. The time I spent with Peter, although short in the grand scheme of my time on the project, was integral to getting a foothold on some of the more stubborn operational adjustments that were desperately needed. A kindred spirit, Peter encouraged me to run with the organizational solutions I was just beginning to bring to the table. With a figurative slap on the back, he gave me the confidence and an open door to push on.

Among the first series of staff changes, I sought to establish a clear structure, a recipe from which the staff could easily implement new daily requirements. This of course involved yet another white board. But before even a daily plan could be erected, some nomenclature was required. We defined each approach to Keiko's conditioning by session types. Sometimes our intent was simply to create exertion, exercise (much of the time early on). Other times the goal might be teaching Keiko something new or simply encouraging him to interact with his environment. More advanced objectives engaged the use of differential reinforcement techniques encompassing all hours of the available clock. Husbandry, or any form of preventative medical evaluation, formed the final category of overt activity from Keiko's trainers.

As we carefully mapped the plan each day, we also meticulously broke down every measurable behavior. The staff was unaccustomed to telling Keiko "no." Almost anything he did before had been accepted, no matter how lackluster the response or how low the jump. At times, even a complete lack of response to their wanting call was shrugged off as if it had no bearing on Keiko's future. Mother Nature would never be so accommodating. I was determined to ensure that every demand on Keiko's energy and responsiveness was met with absolute consistency, no matter who, no matter when. Inconsistency among his trainers would only teach Keiko to discriminate, firstly with the trainer, but later in the context of his environment. So vital was this simple prospect that it kept me glued to the bay pen day or night in supervision of all applied conditioning.

Apart from direct interactions, we began to see the negative space—the time and space between training sessions. In this undiscovered realm, the Behavior Team had to learn to recognize opportunity. They had to open their eyes to any behavior, movement or activity Keiko might engage in that resembled a wild animal. At first it was difficult. Keiko was so accustomed to sitting idle and floating at the surface, we struggled to find any small chance to encourage an active whale. But slight modification to Keiko's diet, increased level of exercise during sessions and our unwavering consistency between trainers finally began to take tangible form. Applied together, these shrewd but simple changes bit by bit began to awaken the animal within.

Once More unto the Breach

Struggling to condition our would-be athlete, we desperately needed more exercise behaviors. Up until this point, we had only to rely on bows (straight dorsal-up jumps from the water), fast swims, tail slaps and a couple versions of a ventral or upside-down swim. We needed more aerial behaviors that would force Keiko to truly exert himself. Shortly after the second rotation arrived, Brian O'Neill, Tracy Karmuza and I began teaching Keiko a side breach, a natural behavior to add to his repertoire.

The introduction of a previously unknown and energetic behavior proved to be beneficial on many levels. By teaching Keiko a new behavior, it stimulated him mentally while also providing a platform for rehearsing the basics of behavioral modification with Brian and Tracy.

Brian was about my age, early thirties, and had been the lead on Keiko's daily care prior to my joining the team. He didn't appreciate the bump in his responsibilities (or freedom), nor did he appreciate taking orders from me. Understandable perhaps, because unlike the first crew, Brian actually had some zoological exposure to killer whales before working with Keiko. He had been in the Animal Care department at SeaWorld of California sometime before joining the Keiko Release Project.

Within the SeaWorld system, there are two distinct departments that interact with the marine mammals in the parks: Animal Care and Animal Training (veterinary care being a part of the former). This dichotomy in animal responsibilities had been the creation of Dr. Lanny Cornell during his zoological dictatorship at SeaWorld. This same separation of animal roles survives within the SeaWorld system to this day. Animal Care principally dealt with all medically related management issues and responded when an animal was stranded on a beach or when a marine mammal in the park was in need of special care or transport to a new location. On the other hand, Animal Training was solely focused on behavior and training. The poorly designed separation of responsibilities effectively created generations of animal professionals, some of whom lent little value to behavioral sciences and others who were never taught much in the way of animal physiology.

In the SeaWorld system, killer whales were always the primary responsibility of the Animal Training department. In Brian's time at SeaWorld, it is likely his only exposure to killer whales would have been in the event of an animal transfer to another SeaWorld park or if one of the whales had been seriously ill. Else he would have little to no interaction with them.

Even so, it was beneficial that Brian had come from the same zoological institution as Robin and I. It meant that we spoke the same professional language. It also meant that Brian possessed a preprogrammed dislike of animal trainers; another useful creation of the "Lanny Cornell School of Marine Mammal Management." Fortunately, Lanny had been out of the SeaWorld system long enough that the old Hatfield-McCoy relationship between the two departments had softened somewhat during Brian's and my tenure there. We were able to find that we had more in common than not, and as a result, we steadily found our way in working together.

Brian was very fit. Despite being a smoker, he worked out religiously and had the coveted genetics that piss off the average guy. He could workout half the time and get twice the results. Having been alike in our shared addiction to physical fitness, our friendship began mostly with discussions about workout routines. But that didn't last, as very quickly Brian and I discovered that we both possessed an uncanny and complimentary ability to goof around. Pretty soon, we were having fun at Tracy's expense on a daily basis. When spending twenty-plus-hour days together, day-in and day-out, it does not take long to get to know a person quite well.

Just as Brian and I were at the peak of our newfound friendship, I was nearing the end of my first rotation. Soon, Robin would return and after a couple days of overlap, I was to head home for a few weeks. Brian, Tracy and I were determined to finish training Keiko's side breach before handing it over to the next crew.

It was another cold and wet day on the bay pen, the third in a row. After the prior week's gorgeous weather, it was difficult to resist the urge to nap all day in the research shack. Regardless, we had a side breach to finish, and Keiko was doing extremely well lately, making it only that much more rewarding for Brian, Tracy and me to endure the bone-chilling dampness.

We were up to almost 100 minutes of exercise a day with Keiko, and he was moving a lot more in his free time. I was pretty happy with the progress we made in May. We had accomplished a lot more than I expected starting out.

Tracy was standing to my right as Brian was preparing the food for the day's first side breach training session. I took the target pole (a long pole with a buoy attached to the end giving Keiko a position and "target" to follow) and looked wide-eyed from Brian to Tracy. Tracy, not being one for subtlety, was never short of a reaction (and Lord knows I didn't need an audience).

"No! I am not going to be the target wench again!" Faking a cry she continued, "It's Brian's turn, and I don't have my splash suit anyway." The bottom lip came out for emphasis.

Smirking and with his whistle in his mouth, Brian had already picked up the fish bucket and was walking away from Tracy and me.

Tracy was so easy. "Mark! I swear . . . Brian, get back here!"

Laughing, I let her off the hook, "Okay, okay—I'll do it. You take the B point at the second position. Brian, what are you going to look for?" This left Brian to work Keiko.

Brian, deflecting a punch in the shoulder from Tracy, had walked back into the small circle. "If he gives us one really good one, let's end it there and rub him down with the brushes," he said while chewing the end of his whistle bridge.

"I like it," I said, hoping it would be our lucky day.

In the past few sessions Keiko's execution had been solid, but now we needed that *umph*, that energetic push up and out of the water that would allow him to slap his side down hard on the surface—giving us a true side breach. I didn't mind the imminent splash of Keiko's two-ton landing. If he did it right, I'd gladly be the sacrificial target "wench," splash suit or not. To get the breakthrough the target work would have to be solid. I had always excelled at target work, and I liked doing it. *Once more unto the breach!*

Brian stepped up to the north pool and called Keiko to the west side of the octagon. As soon as we saw Keiko respond to Brian's call-over, Tracy and I ran to our positions. I was almost opposite Brian, and Tracy stationed herself another quarter way around the circle to my right. After Brian gave Keiko the signal, I would use the sixteen-foot long pole with the small buoy on the end to slap the surface of the water letting Keiko know what we wanted him to do. If he got it right, he would exit the water with his back (dorsal fin) toward me, chasing the target on the end of the pole. At the height of his jump, I would snap the target down to my right, and Keiko would "snap" his head over, putting his body in perfect side breach position.

The snap was the magic ingredient. All our previous approximations focused largely on Keiko responding tenaciously to the movement of the target. (An approximation describes the process of breaking complex behaviors into smaller, easier steps or "approximations.")

The intensity in his following of the target is what turns an otherwise off-centered spyhop into a full lateral side-breach. (A spyhop is vertical half-rise out of the water performed by a whale, normally to view his surroundings.)

Tracy's job was to slap the water ahead of Keiko as soon as he completed the movement. By directing him to continue in the same counterclockwise path, Tracy was ensuring that Keiko would maintain a "laid-out" position on reentry, the follow-through. It's analogous to a high diver looking toward where he or she is headed on a standing back flip or a somersault high dive.

Brian assessed Keiko. If Keiko looked uninterested, he would have ended the session. In order to get the breakthrough we needed, Keiko had to be alert and energetic. It was one of the reasons that we chose first thing in the morning to "go for it."

"Brian, if we get it, run over to Tracy as soon as you hear my bridge, so we can all three brush and feed him!" I had to yell, even across the relatively short distance of the north pool; the shifting wind could easily steal my words. Brian's nod was barely perceptible. He didn't look at me, but I knew he heard me. I also knew he didn't like the "chafe," but if I didn't repeat the obvious, he would just as often forget and stand there as if we hadn't just talked about it. This was a difficulty I always had with Brian, when he was "on" he was great and an absolute blast to work alongside. But when he was off, there was no penetrating that glum poker face or "Eeyorelike" attitude toward life and everything about it.

Today it appeared that both Brian and Keiko were ready. *Come on Big Man, sock it to me,* I thought. As if reading my mind, Tracy's giddy chuckles to my right reminded me of her firsthand experience as the target wench.

"You pipe down over there, missy, and be ready to slap if I bridge," I chided. Tracy was somewhat younger than the rest of the staff and constituted the token female affection of almost every guy on the project. She had long dark hair and was very attractive in all regards. I rather enjoyed teasing her and usually that meant being boldly rude.

"You miss this, and I'm exiling you to the top of the research shack for the rest of the day," I yelled while watching Brian and Keiko. He was ready to send the breach.

"Bring it, Mr. Party Boots," Tracy yelled back. As Keiko was leaving Brian's position with urgency, I thought, *Damn, this looks good.* Then Tracy's comment resonated, reminding me of how full of forty-degree water her boots usually got from even the earlier rehearsals on side breach. *Too late now.*

I slapped and waited with the target held low over the water looking for that exact moment, when he was on his upward run and focused on the target buoy. Then quickly raising the target high over my head I turned to my right ready to move horizontally with him on the snap. He was up, *damn he was up*—in that splitsecond of thought that renders everything in slow motion, I realized *this was it—he had it—this was going to be huge*. I snapped the target to the right and down to the surface while moving with Keiko. It looked almost like he had cleared his dorsal fin out of the water . . . by a lot. I thought I might have even seen flukes near the surface.

There was that oh-so-familiar thundering "smack" just before the splash knocked me into the bay pen handrail. That was only the first.

The second-wave splash made sure my boots were filled to the rim. *Holy crap, that is cold water!* I didn't care, all I could hear was Tracy yelling at the top of her lungs and making a commotion. Even Brian was whoop'n it up. This was definitely fun—the Big Man got it. His first full-on side breach, and it was a good one. Wish I had seen it. All I saw was a wall of water.

As I emptied my boots of the cold water, Keiko was just sitting up in front of Brian and Tracy on the north end of the pool. Like a child full of newness, he sat high in the water, almost expectant; his eyes wide and mouth gaping. Tracy shoveled a few heaping fistfuls of herring into his mouth as she maintained her high-pitched praise. It was garble to Keiko, but he understood the energy. I joined them and the three of us began scrubbing Keiko with floor brushes reserved for just this purpose. Keiko happily obliged, rolling and presenting first one pectoral flipper then the other and then rolling upside down lest we forget his belly.

He had nailed the side breach perfectly; high marks on both energy and execution. But more enticing was the clarity of the message given and received. There was no doubt in my mind; Keiko wouldn't forget this morning's success.

We just started work on the side breach less than two weeks before. Yeah, I'll take that any day, and twice on Sundays. This is where it's at . . . the heart of working with animals. To present something new, give them the conditions and the clarity to succeed, step by step . . . and then to witness the breakthrough and even more stirring, the animal's recognition that they've got it. It is better than any artificial high I've ever known.

E-mail: May 20, 1999

> To: Alyssa (and family)
>
> Subj: Pictures of Meeeeeeee (oh . . . and Keiko)
>
> Couple pictures of side breach training attached.
>
> The pictures really don't capture the true nature of the environment . . . this is me, Brian and Tracy working a side breach today (initial stages) in a hail storm, 34 degrees, wind at 30ish with gusts to 40 mph . . . if you can imagine about 4 foot seas with a long wavelength moving through the pen then you start to get an idea of what it's like to work the target pole accurately and keep yourself out of the 44 degree water at the same time.
>
> Yeah . . . and we have to walk uphill in three feet of snow to get to the fish house (both ways)
>
> Love to all, Mark

Plenty of challenges surfaced in May and June of 1999; however, for the most part this period produced positive results in Keiko. He was no longer the sluggish, overstuffed and lazy whale I had first met. In place of floating by his blue ball or scouring the perimeter of the bay pen seeking any morsel of attention, now he swam more and solicited less. And although he wasn't yet the lean, mean survival machine we aimed to create, slight improvements in his physique were beginning to emerge. Watching his responses to the staff and increasing bouts of alertness, I was convinced we were beginning to see a Keiko that had never before existed. Truth be told, I was not completely satisfied with the progress; I knew we could move faster. It was only the tip of the iceberg. In classic fashion I was just finding the rhythm, and it was time for my rotation home. Though a great part of me wanted to remain and focus on Keiko's embryonic transition, the break was equally important for maintaining my own well-being.

Having married only five months prior to arriving in Iceland, I was brimming with anticipation at being reunited with my beautiful bride. But before I could depart the island, I had to bring Robin up to speed, covering all that transpired and my thoughts for Keiko in the progression of his exercise program.

Upon Robin's arrival, he and I spent our entire two-day overlap talking mostly about staff roles and changes to the rotation schedules. Much of our time was dominated with growing dissention within the ranks; specifically Jeff's chosen few who routinely maintained the same schedule as Jeff. Whether they were actively in Iceland or home on their off-schedule, e-mails were flying, and Charles made us aware of the sentiments being exchanged among the staff. He wanted the new additions to the release team to succeed and took every opportunity he could to smooth out the wrinkles. By giving Robin a "heads up" on the most glaring undercurrents running afoul within the staff, Charles hoped that Robin and I could selectively ease up on some of the more difficult changes and thus lessen the complaints.

But it was not to be. No matter how hard we tried, there were those on staff that wanted nothing to do with our management of the operation. Specifically, there were those who believed we were pushing Keiko too hard. No amount of education on the science of behavior or the finer points of the release plan changed some opinions. It was a cancer within the ranks of the release team that would eventually have to be cut out.

Eternal Daylight

I returned to Iceland on my second rotation after only three short weeks at home. It felt as if I'd never left. If there was any silver lining, it was that Alyssa and I were trapped in the extended twilight of our honeymoon and each short rotation home rekindled the excitement of our recent marriage.

I had only just started on the project in April and here it was already teetering on the brink of July. It was full-on summer in Iceland. One unmistakable characteristic of this far-north summer is that the sun never truly sets. By the middle of that next month, direct sunlight kept us company twenty-three hours a day. There were only a few minutes each evening where the sun momentarily dipped below the horizon, dimming the otherwise eternal daylight. The never-ending day was operationally beneficial, allowing us to make much progress in pursuing our goals with Keiko. By extending and alternating our shifts on the bay pen, we could focus attention on Keiko's behavior around the clock.

It also meant the staff would not get much sleep. Evening festivities (late-night drinking binges) routinely bled well into the next day, rendering the occasional team member incapable of making early morning shifts on the bay pen. Looking back, I suspect that some of the increase in extracurricular activity might have been avoidance of Robin and me. We had taken all the comfort out of being on the bay pen. Being with or around Keiko now required thinking and work. The days of sitting on the pen dabbling on the computer or sleeping off the previous night's festivities were gone. During this period in the summer of 1999, many staff members became increasingly apathetic toward working with Keiko although I did not see it this way at the time. I attributed the disinterest to a historically lax work environment and a direct challenge to what Robin and I were implementing. I suppose a little of each was true.

Though symptoms varied, at the heart of the matter was the simple human resistance to change. A select few felt as if they had been pushed aside, their authority diminished and their opinions worthless. That Robin or I maintained a constant presence on the bay pen was perceived as distrust in the staff. In reality, close supervision was necessary to ensure that the exacting requirements of Keiko's program not suffer from inexperienced application and, with some, the outright inability to comprehend various elements of the rehabilitation process.

Behavioral conditioning requires consistency, which in turn requires patience. Like many novice animal trainers, the lack of immediate results often led to open season on suggesting changes in approach, many of which were based on an emotional need to coddle Keiko. Try as I might, many attempts at explanation seemed wasted on unwilling ears. The drawn-out effort to educate did little more than fuel my growing impatience. I was hell-bent on investing our collective time in forward progress and the never-ending need to validate every component of Keiko's daily plan was exhausting. I often shared these frustrations with Alyssa, who always talked me back from the brink of disaster. She reminded me that lacking other means to contribute, the staff's affection and commitment to Keiko would materialize in other ways, ways that I was too quick to accept as belligerent resistance.

Although I took her guidance to heart, isolating the person from the problem, the merging of multiple tiers of incompetence surrounding the project would at times bury me alive in trepidation. Nonetheless, had we been left to our own in Iceland it would have been almost easy. Instead, Robin and I often spent as much time educating the FWKF board through our interactions with Charles as we did rolling up our sleeves and guiding the more important work with Keiko himself.

Within days of my return, a board member visited the operation. During a short spell on the bay pen, the board member proceeded to "talk" to Keiko as if she were having a conversation. At best it provided nothing more than novel material for short-lived levity, until I learned the person believed that Keiko was a member of an Intergalactic Cetacean Spaceship and that whales were here on earth to plea to humans for better treatment of our world. Beyond the sinking feeling in my stomach, I became intimately aware of how unrelenting the battle of continual education would persist throughout the organization. This, of course, placed the challenges on-site in a new light and if nothing else made them seem trivial by comparison.

Lundinn

Whether or not escape was a motivator, there were ample reasons for the staff's allure with experiencing nightlife in the small village. The town of Heimaey is a vastly intriguing place in which to socialize and the cultural "aggressiveness" of the Icelandic people is just too good to miss. I'm not talking about "fight-night" at the local pub. By aggressiveness I mean that Icelandic men and women are very outgoing with their social affections.

On one of my first forays into island nightlife, I was asked to dance by a local Icelandic woman, repeatedly. It happened in the nearby watering hole, Lundinn. A pub half submerged beneath a two- or three-story building, the quaint and somewhat rustic interior along with the small floor space, imparted a very cozy atmosphere conducive to meeting new people. An attractive woman sat across from me at one of the tables lining the dance floor. She was in her early thirties and her blonde hair, blue eyes and thin figure placed her right in the middle of the bell-curve of Icelandic women. Iceland has the most beautiful people congregated in one land I've ever witnessed.

"You here with Keiko whale?" she said, making Keiko (cake-o) sound more like Keeko (ceek-o). Icelandic accents are very similar to native German accents when speaking English. The guttural pronunciations translated to hard consonants. She skipped the occasional pronoun and applied unique interpretations of the softer vowels in English.

"Yes, I am," I replied simply.

"You like Iceland?" she continued.

I could tell she had been drinking heavily, but her happy smile elicited the same in me. I felt a little silly really. I knew she was flirting, but I wasn't looking for anything in that department. "I love it," I said, trying hard not to encourage her.

"We dance, yes?" But it wasn't a question. She said it as she was grabbing my hand and lifting herself from the chair.

"Oh no, thanks. I'm not a dancer. I'm just enjoying the music and watching my friends there playing the slot machines," I replied while resisting her pull on my arm.

She sat back down. "You're new with Keiko, right?" she asked. Doubtful any locals on the island did not know the project staff well by now.

"I started in April, but this is the first time I've been to Lundinn."

Stephen Claussen had seen me talking to her from across the room. Standing by the slot machines, he gave me a wink and a suggestive smile. Now I felt really silly.

"Come on, I show you good dance. You will like it," she pushed. She never released my hand.

"No really, I'm sorry but I'm just not into dancing." This was going nowhere fast.

Before I knew it, she had pulled her blouse down just above the most private parts of her breasts. "You don't want to dance with me because my breasts not big enough."

A rather innocent prude, I was shocked and let it show on my face all too easily. She laughed. "No, of course not," I shot back. "I'm a terrible dancer, I just don't like dancing. There's nothing wrong with your breasts."

At that, she feigned a small pout, touched my cheek with the back of her hand in a caressing manner, smiled warmly and left the table. I felt like a schmuck, but a safe schmuck.

Later that evening as the few of us from the project were leaving, we stood around in a semicircle waiting for the last conversations to wrap up. The woman returned and placed herself next to me and was holding my hand. She stood uncomfortably close, leaning into me with her shoulder. On her right was a man I hadn't seen before. She was holding his hand also.

Walking out of Lundinn I shared, almost confessed, what had happened with our Icelandic host (a coworker on the project) and asked him who the man had been. He informed me that it was her husband. I couldn't contain my incredulity. My first instinct was to fear conflict with this unknown man over my interactions with his assertive wife. Mercifully I was let off the hook pretty quick as everyone in our small group had a good laugh and explained that this was normal behavior for Icelanders. They told me not to worry, and that many Icelandic couples stray. In an almost nonchalant manner, they described the ritual of many spouses walking back to their homes in the early mornings following an adventurous night in another's bed. This was hard to understand, especially for a prudish newlywed. Although not a customary norm in Icelandic society, I would learn that this practice was certainly not uncommon, at least not on Heimaey.

Lundi Pysja

Nightlife in the small town was far from the only novel attraction that offered a brief escape from the pressures of the project. Klettsvik Bay is a colossal bird sanctuary. There, an unusual aerial display often commanded our attention. Among the many species of birds that frequent the bay, one of the most memorable is the puffin. Small seafaring birds, puffins almost looked phony, like plastic children's toys. Their distinctive black and white coloration appears as the avian version of a killer whale's disruptive camouflage. Puffin, or "lundi" in Icelandic, are black on their backs and wings and white on their breast and underside. The two colors are divided perfectly in clean lines, giving them a man-made appearance. Contrasting with the simplicity of this design is a very colorful orange-white-black striated bill. They have larger heads than appear proportional to their little bodies. The puffin's eyes are framed by black triangles that makes them seem "concerned" or "sympathetic." In flight, they are fast and hyper, legs trailing to the sides of their short butts and jerking to and fro, making their way through the aerial mob over Klettsvik. During the spring season, many of us took great pleasure watching lundi flying and diving throughout the bay.

I loved watching the puffins landing on the water. They zoomed in, full of confidence as if little airborne mavericks and just as they reached the surface on a long, low trajectory, the feet went out like landing gear catching the water at high speed and sending them tumbling across the surface. It was one of the funniest things I had ever witnessed in the animal kingdom. Those hysterical landings never became commonplace. We always stopped to watch the puffin's signature "crash-and-burn" style landing.

Perhaps a less popular scene, but comical nonetheless when witnessed from afar was puffin hunting. High up on the grassy tops of the Klettsvik cliffs, puffin hunters would hide behind large rocks or unusually sizeable grassy knolls. Huntsmen used long poles with hoop-nets on the end. We could never see the hunters, at least not until they lurched up and netted the unsuspecting puffin right out of mid-flight. Viewed from hundreds of feet below on the bay pen, it looked like an aerial version of "whack-a-mole."

In August a fascinating event, aptly titled the "Puffin Patrol," takes place on Heimaey. Young puffin, called "lundi pysja" (LOON dih PIHS-yah) leave their nests, holes high in the hilltops and cliffs surrounding the island. At night, the lights of the town attract them, and they glide into the streets, yards and gardens of Heimaey by the hundreds. Townspeople allow their children to stay out late in August to collect the little lundi pysja into cardboard boxes, shoe boxes or anything that will suffice for the short visit. Some children might collect as many as ten or more young puffins in one night. The next morning in the daylight, the children take their catch to the seashore where the puffins are then thrown high into the air, gliding off toward the sea and back to their intended destination. It is a very charming tradition to witness and one that deserves a place on the bucket list.

E-mail: August 11, 1999

> Subj: Had to go
>
> To: Alyssa
>
> *Here is one of the perks of the job . . . last night we had an Italian millionaire (who had donated to the project) tour the facility and watch a training session with Keiko. They then went out and watched wild whales. It was a beautiful day and around 7–8 o'clock in the evening the sun was spectacular on the Iceland glacier and surrounding mountains. The millionaire invited us to dinner on his 250-foot private ship. I have never seen anything like it. It is impossible to describe the amount of money this guy must have. The ship was everything you can imagine and more than I can tell . . . endless teak, brass, stainless steel, helicopter pad, two 40-foot tender boats on deck (with a crane lift to put them in the water) . . . six jet skis, four ATVs (four-wheelers), bicycles, two motorcycles, ocean kayaks, a dive room with all the equipment you could ever wish for. I will have pictures to send but they will not be digital so it will take a little longer. It makes you realize that nonprofits are the benefactors of "extreme profit."*
>
> *Love you,*
>
> *Mark*

A Million Dollar Solution

July and August produced many more questions than answers about how the project would move forward. First we needed to consider how we would physically get Keiko to the open ocean. After all, we couldn't just open the bay pen gate and hope for the best. Our location well inside of the shipping channel, and the proximity of a harbor bustling with human activity presented far too many variables. Lanny's idea of airlifting Keiko to the first available pod of wild whales and dropping him in the ocean was analogous to tossing a family pet out the car door on a Sunday afternoon drive through the wilderness. Probably worse. Fortunately, there wasn't a soul on-site that would give heed to such a farcical concept of release.

This left us with little other means by which we could guide Keiko in the right direction. Charles, Robin, Jeff, and I agreed that Keiko would be escorted to the open ocean by training him to follow a special-purpose "walk-boat." In other words, we would teach him to follow alongside a boat much like a dog trained to heel at its owner's side without a leash.

A few essentials would have to be addressed to make the designated walk-boat stand apart, but of immediate concern was when and where to introduce a walk-boat. It wouldn't do much good to put the boat inside the bay pen and for reasons stated, we couldn't take the risk of bringing Keiko out of the pen on blind faith that he would follow the boat. That was a behavior, and like any other behavior, Keiko would have to learn how to follow the walk-boat.

Beyond our fears of what Keiko might or might not do once outside the bay pen, the prospect of having access to the open ocean also meant that every criteria required for release would have to be met beforehand. After all, a permit was required to release the whale. We couldn't just wake up one morning and decide to free Keiko. Exposing him to the open ocean was by any definition a "release" scenario, even though the intent at this stage was only to train Keiko on heeling alongside a designated walk-boat. Obtaining permission to take him out of the pen would be predicated on evidence that Keiko had met the prerequisites for release. These restrictions were common sense. They were intended not only to protect Keiko, but also humans and the indigenous whales that migrated around Vestmannaeyjar. To allow Keiko access to the wild before he was actually ready would be nothing less than unadulterated negligence.

Exercise was certainly making improvements in his physical condition and by midsummer, he was routinely completing over a hundred minutes of strenuous exercise per day. Keiko was even looking much more alert and responsive than ever before. But despite any encouragement afforded by his improving physical prowess, we couldn't just go straight from the bay pen to sending him off to sea. Yet it was impossible to address all the elements required for official release from inside the relatively small bay pen.

Although masterful in design and function, the floating facility was limiting. We could keep Keiko fit and mentally stimulated in the bay pen, though only through continued human interaction and enrichment. It was time to increase his activity beyond training sessions. Likewise, it was time to begin reducing his interest in human activity.

How could we prepare Keiko to meet the challenges required for release without the walk-boat? How could we train the use of a walk-boat without taking Keiko out of the confines of the bay pen? At the crossroads of necessity and constraint, our thoughts and discussions focused on an interim step.

The million dollar solution was to give Keiko a bigger and more varied environment. To surmount the next hurdle in his move toward unrestricted ocean access, we would build a barrier across the mouth of Klettsvik Bay. In so doing, we could substantially increase Keiko's habitat from the comparatively restrictive pen to the grand expanse of Klettsvik Bay. A gargantuan net would grant us the tool to prepare Keiko further. The solution would simultaneously allow us to develop each necessary component of Keiko's rehabilitation, including the walk-boat, while also delaying the need for a final release permit. It would have been a simple solution, but for the raging currents and winds of the bay.

Constructing an 800-foot-long by fifty- to sixty-foot deep net across the mouth of a bay in the North Atlantic is no small feat as evidenced by the fact that the feasibility of such an undertaking was initially deemed *impossible* by more than a few engineers. Adding perspective, at least one of those engineers was German. As a Westerner with a healthy respect for German engineering, the early verdict issued in response to the barrier net concept was sobering. Not to be undone, our gallant Marine Operations crew pushed forward. If there were any chance that such an undertaking could be accomplished, they would find a way.

One lone engineer within Woods Hole Oceanographic Institute found the idea plausible. Without delay, Robin traveled to Massachusetts and met with the Woods Hole team. By the end of their weeklong collaboration, they produced a preliminary concept design for the barrier net and provided the project with a clear path forward. With September upon us and the foul weather of winter just around the corner, an urgency to solve the most stubborn impediments in the barrier's design and practical application began to dominate everything and everyone.

Growing Pain

Prior to our (Robin's or my) involvement in the project, the management of Keiko's release had largely been a grassroots effort. The original cast was more a small group of friends than any semblance of professional organization or experienced management, at least on the merits of a never-before-attempted killer whale release.

The youth and inexperience of the staff combined with a "let's party" atmosphere gave way to a highly emotional undercurrent that plagued the operation following the FWKF's merger with OFS. Ocean Futures Society had only become involved shortly before Robin was initially contacted. We (OFS included) quite literally walked into a volatile situation. To the less experienced staff members on-site, we were outsiders; strangers who had no right to invade their home and crash their party. In many ways the group acted as if Keiko belonged to them alone. It was as if we were taking control of their family pet. This possessiveness was not uncommon in the realm of the animal field. Anytime a person or group of people dedicate their time to the care of any animal, an emotional sense of ownership slowly but surely roots itself deeply within those caregivers. In almost every case, this by-product creates an unhealthy barrier to making decisions that are in the best interests of the animal. They had a perception of killer whales fashioned by their interactions with this one animal (a highly unusual male killer whale to begin with) and believed no one knew his particular needs better than they.

Every decision had been emotionally based with no program or plan in place to guide them. It didn't matter how well we communicated or how much we communicated with regard to the release plan we had been hired to implement. To them, everything we did was offensive. *Who were we to come charging in taking control of their whale and telling them what he should be like?* The fact that this emotional and territorial attitude even existed on a high-profile, high-stakes project such as this troubled me greatly. From where I sat, they defiled the throne of *Orcinus* orca by making Keiko their private playmate. In effect, they minimized the ocean's top predator into a completely lethargic and obese perversion of a killer whale. How they imagined their approach was preparing Keiko for the harshest life he would ever know, I could not and would never come to comprehend.

It simply didn't matter to them that Robin and I had worked with more than twenty-eight different killer whales. They felt knowing this one whale afforded them a more valid foundation. Keiko became their entire life, and that was precisely the problem. Their vision of Keiko held hostage any real potential he might have had toward independence. In essence, they disregarded the challenges that faced Keiko, instead focusing on their bond with him. Although we did everything we could to assuage the more stubborn trepidations, we could not allow the "ownership" sentimentality over Keiko to prevail. In one of many such instances, the staff's desire to spoil Keiko created direct conflict.

It was late summer and we had finally approximated the blue Boomer Ball out of Keiko's environment, effectively purging his hopeless love affair with an inanimate object. It was a slow process and much work had gone into reaching this important step. Yet upon my return to Heimaey, on my third rotation, I came back only to find the Boomer Ball once again floating in the pen with Keiko. After a condensed repeat of the same gradual withdrawal, I promptly removed the toy and cut it into pieces lest his dependency on the ball be reversed again. The Boomer Ball incident and other perceived conflicts only heightened tensions between us. There was little wiggle room left in any resolution. We would not and could not continue to allow the discord to reach Keiko or disrupt what tenuous progress existed.

Karen McRea, in particular, had been willing to fight us at every turn. Karen was one of Jeff's favored few. She was a popular member of the release team, not only with the staff, but also among select members of the FWKF board. Emboldened by youth, she fought to protect what she believed was best for Keiko. What she disliked of our management of Keiko she made well-known through informal channels. Karen had become the de facto spokesperson championing Keiko's defense. At the same time, she represented what disadvantaged Keiko most.

Like growing up, the project had to be matured into a new entity, likewise, that meant growing pains. We could no longer drag along the most rebellious member of the team. After extensive talks among Charles, Robin, and Jeff, it was decided that OFS would not renew Karen's contract, effectively removing her from the project.

Robin had agreed to take the responsibility to address Karen, after all it was our assertion that it must be done and therefore we had to own it. Unfortunately, we tripped over our own feet and at least temporarily caused more harm than good. Robin's schedule placed him off-site just as Karen's contract came due for extension. By default, the task fell to Jeff, the most beloved leader of the original teams. In the aftermath, the incident did no favors for

Robin's and my relationship with Jeff. The staff considered the act cowardly on our part. We pushed for Karen's termination, but by outward appearance were unwilling to carry out the order. It was a mistake that cost us dearly in the bank of human equity.

Karen's removal from the project paralleled other changes that had to be made. It was time to break up the party. We began reorganizing the rotational schedules, instead electing for "rotational roles." Individuals rather than entire teams were rotated. This staggered shift changes so that only one person was "new" at any given time. This change effectively eliminated the two-team division, requiring many staff members to work with counterparts they were unaccustomed to seeing often. The transition was rocky to say the least. It was not made any easier by Karen's dismissal from the project and her lingering interference at the organizational levels. After she returned to the States, the popular Karen continued to communicate with team members, the FWKF board of directors and even donors who had contributed to the project.

While I could only guess what her goal might have been in those communications, all that is apparent is what resulted. Charles became so inundated by criticism from the board that he ordered no one else could be forcibly removed from the project again. It was an absurd decree and one only a nonprofit could possibly uphold. No self-respecting company in the free world would allow a single employee to hijack or threaten the organization's mission. Yet this is exactly what OFS and the FWKF board were doing. Poetically, just as the staff viewed Keiko the object of their affection and protection, so too did the board view the staff, and regardless of performance.

The "decree" removed any teeth from real accountability, which directly led to an atmosphere furthering feigned cooperation. Whenever Robin or I were not individually on the bay pen overseeing Keiko's program, the staff did whatever it felt like doing. In most cases that meant breaking protocol, even, at times, getting back into the water with Keiko despite that fact that Keiko's rehabilitation had moved beyond the intimate human association. The pendulum swing of inconsistency in Keiko's interactions only served to create noticeable setbacks in his progress.

One month he would surprise us with ever-increasing activity, much-improved energy level and the overall appearance of the alert animal we were hoping to discover. The next we would see an increase in thrashing behavior and logging at the surface. There were many other signs, obvious changes in Keiko's behavior that provided ample evidence that program guidelines were not being followed. Robin and I tried to make certain that at least one of us was always on the bay pen; however, this would not be practical for long. Robin had many other responsibilities within the project so it was not always feasible for him to allocate his time solely to the bay pen. Only one solution remained, we needed to find additional experienced behaviorists to help maintain consistency in Keiko's rehabilitation.

E-mail: August 13, 1999

> *Subj: hello, my love*
>
> To: Alyssa
>
> *Tried to call you, but I cannot get a phone with privacy (unless of course I stay up late enough that everyone else has gone to bed—which is usually later than I can handle). There is one phone—it's in the office—where everyone has to go to get to the computers—which are always occupied—because it's the only way to communicate with your family. Anyway, I can't seem to get a grip on things lately. I have had great success with Keiko, because I have been working with him myself 95% of the time. Other than Keiko, I am not enjoying this anymore and it makes it difficult to maintain focus with people, and making employment decisions. I have a hard time considering bringing someone onboard when I feel tired of the whole project, the people and the poor decision management system. I am also sick of letting things fall apart when I can't be there to baby-sit. I am tired of everything about this project dominating my life, and especially the fact that it has put you and me on ice.*

> *I am sorry to dump . . . I intended on writing a sweet e-mail when I started . . . again, it just controls my every thought.*
>
> *I do love and miss you more than I can possibly describe . . . and need your counsel, too.*
>
> *Mark*

Adrenaline

The magnitude of challenge exacted by the unpredictable weather of the North Atlantic confounds accurate description. This is not to say that every waking moment was fought into the wind. There were intervals of mild weather and at times, even spring-like sun producing greatly cherished, crisp beautiful days. But when the weather surrounding Heimaey reared its ugly head, even the simple task of standing upright became a struggle.

Klettsvik Bay was in most ways an ecosystem unto its own, down to a climate often juxtaposed with that of nearby downtown. On many an occasion, we would leave the pen, exhausted from the constant barrage of vertical water and pounding wind, only to find the town basking in sun with a pleasant breeze. Without question, the turbo-scoop characteristics of Klettsvik angrily amplified every gift the Gulf Stream would send its way; winds often in excess of 100 mph and in many instances maxing out the upper limits of our anemometer gauge with prolonged gusts over 175 mph.

In the first weeks and months of my time on the project, my fellow expatriates recounted plentiful and colorful stories—most all of them set in a weather-related plot. On one occasion, Stephen Claussen had lost his hat, blown from his head while standing outside the research shack on the bay pen. The hat was sucked hopelessly upward spiraling into the aerial abyss, only to be returned to the same vicinity nearly forty-five minutes later. Like this one, the tales were difficult to accept, and on first hearing them, it was natural to assume they were dramatically exaggerated. I now contend that stories involving the weather in Klettsvik cannot be overstated.

Nothing could be taken for granted. Nothing could be left out on the deck and every container or locker or storage bin on the bay pen had to be expertly lashed down via Texas trailer-hitch knots or secured with ratchet strap or chain, or else the contents and the container would become hazardous projectiles. Those of us who practically lived on the bay pen attending to Keiko's daily program became hopelessly addicted to adrenaline. In some of the more docile winds gusting between 101 and 140 mph, we would don our splash suits, jump on the portable Jet Ski dock in the medical pool and see how long we could hold on before we were either bounced into the pool or our arms and hands fatigued. When assaulted by the more serious winds in excess of this range, we could hold onto the railing of the bay pen, watch for a wall of water carried by a sizable gust coming off the west rock face and at just the right moment, give a little jump. If timed right, we could "Superman" for a few brief moments, body and feet suspended horizontally in the wind. The more talented among us held records close to three seconds. Most difficult in this insane practice was holding onto the rail, which was a large diameter and far from conducive to the task.

I recall many an exercise session with Keiko whereby the orchestrator of the session had to be fastened to the pen's rail by a safety harness. Without it, we couldn't even free our hands to give signal to Keiko. Just as ridiculous were the varied contraptions we never perfected but which were intended to secure the thirty-five-pound fish bucket so that we could feed Keiko amidst the conditions. Even trying to place fish in his mouth, only inches from his teeth, the occasional herring would be spirited away. Well acclimated to the conditions, Keiko wouldn't even go after the flying fish; he had learned it was a wasted effort.

Still, there remain a handful of storms that were chief among the many we encountered. In one such instance, Jeff, and Steve Sinelli were on the pen together. Jeff, like many of us after months acclimating in the extreme conditions, sought thrill rather than to sit idle within the safe and monotonous confines of the research shack. It was a particularly intense morning.

As Jeff later recounted what happened, he had made an excuse to inspect the dive locker and eastern extents of the pen moorings, requiring a trek of approximately seventy feet up and over the bridge joining the two sides of the odd vessel. Steve was required to "spot" Jeff from the lee side of the research shack, to make sure someone would know what had happened if Jeff disappeared. Steve, with a menacing grin, decided to videotape the excursion.

Pinning himself firmly against the northern lee side of the research structure, Steve stabilized the camera and himself by pushing into the outside wall. Adorned in his Mustang survival suit, Jeff set out for the bridge. He waited for the precise moment between the more aggressive gusts to leave the protection of the shack and make for the bridge. Jeff covered the distance of approximately twenty feet in just a few clumsy leaping steps supercharged by the wind at his back and then careened into the bridge handrail, his ribs taking the brunt of the landing.

The top handrail firmly nestled in his armpit, he made his way to the top steps and began inching across the expanse of the bridge. Turning his back to the unforgiving gusts, he traversed the structure crouched as if he were sitting in an invisible chair. The hood of his Mustang suit had become a mini-amphitheater, the white noise of the wind roaring in his ears as he covered the distance and made his way to the dive locker.

After short inspection, he fought his way back to the bridge, this time leaning at a forty-five degree angle and into the shotgun gusting wind. Eying his destination, Sinelli was nowhere to be found. *Ya fucker*, Jeff thought. *He's supposed to be watching my back.* Finally making his way back across, Jeff entered the research shack, the decibel levels leaving his ears ringing in the relative silence of the interior. There before him on the floor was Steve in a pool of blood, video camera strewn to the side and still running.

Shortly after Steve recovered from the incident, the two reviewed the videotape. Clearly evident in the footage, Steve had been picked up physically, floated for the briefest of moments and then violently thrown almost twenty feet and into the lower steps of the bridge. The impact drove his shoulder into the Chemgrate while his head hit the first aluminum step. Through the static grind of the audio a perceptible *ugghh* emphasized the force of Steve's crash to the ground. Stumbling back to the research shack, he had made his way inside and collapsed on the floor, never having lost grip on the camera. Although Steve had been in a theoretical protected location behind the shack, there was in fact no true protection from the wind that bounced around the bay like bullets off rock. This was just one of many otherwise unbelievable experiences brought to life by the unpredictable ricocheting winds inside Klettsvik.

Staff exchanges on the bay pen constituted harrowing battles with the elements, too numerous to share, as transfers to and from the harbor were many times each day. I gained invaluable boat-handling experience compliments of the extreme conditions, as did everyone on the project. Docking the boat on the bay pen required keen skills at maneuvering and timing. By default, we always docked on the lee side of the pen driving into the wind. On a forty-five degree angle of approach, we motored into the pen, at the last moment simultaneously turning the boat's propulsion directly at the pen and reversing. When well executed, the forward momentum was offset by the reverse action and delicately presented the beam of the boat alongside the pen. The method was simple physics made exasperating by the erratic gusting of the wind. One second it was pushing the boat back away from the pen requiring increased throttle, then only to drop at just the right moment sending the bow careening into the pen's superstructure.

Some pilots were more adept than others and like many things, some pined for the challenge while others welcomed a replacement. In the more moderate to high winds, only Greg or Michael would captain the transfer and almost exclusively in the *Heppin*. Anything over sixty-five mph was too much for the light weight *Sili* to handle, no matter the captain's skill. That said, there were also plenty of occasions where even the most seaworthy of our transport boats could not make the short journey safely—this included the Icelandic Coast Guard's vessel *Thor*. During these supremely foul days and nights, unfortunate souls on the bay pen were at times left to fend for themselves until Mother Nature once again allowed passage.

Even so, Iceland is not unlike any other earthbound continent. Here we were also blessed with seasons of milder weather. As the long dark days of winter edged toward summer and seemingly eternal daylight, workable weather became more frequent and an extremely valuable commodity. Summer was our chance to make good on many things, from Keiko's conditioning goals to bay pen repairs to the ultimate challenge of erecting the barrier net. All of this was painted with the urgency to exploit the mild weather-window that would close on us again all too quickly.

Mr. Iceland

Many colorful characters frequented the Keiko release effort. Beyond the affable locals who we grew to know so well, the project had no small compliment of native Icelandic staff. Among them was an unforgettable couple, Smári and Lina. Smári Harðarson, was a former "Mr. Iceland," an imposing six-foot-tall Viking with shoulders wide enough to seat two adults and muscles enough to easily carry them.

Sigurlína, or "Lina" as we called her, represented the more refined member of the soon-to-be Harðarson family. She, like so many native Icelanders, was blonde and fair skinned. A good bit shorter than Smari, Lina commanded his world nonetheless. The Harðarson family was a fitness family all around. On the feminine scale, Lina's physique paralleled that of the remarkable Mr. Iceland. We didn't interact with Lina on a daily basis, but she was always a welcome replacement for Smari on the occasional bay pen assignment.

The Harðarsons ran a small company that provided the security detail for Keiko's bay pen facility. Well known and respected in the community, Smari's position as the security provider was all that was needed to deter any local's thoughts of tampering with the operation. Smari himself did not cover much of the security detail on the pen. His brute strength and certification as a commercial diver were too valuable not to utilize in marine operations. In every case where the pen was under threat of destruction by the wind and surge currents, Smari was the brawn and the experience to deal with whatever was needed, regardless of the conditions.

Following a particularly menacing storm, I'll never forget watching him complete his maintenance dive to inspect the pen's anchor system. The ledge was easily one-and-a-half feet, if not two, above the water's surface. After completing his dives, Smari pulled himself out of the water, fully clad in dry suit, roughly fifty pounds of lead weight around his waist, two dive tanks on his back and any other equipment he routinely carried. Smari was the only one we knew who could accomplish this feat.

He had quite the sense of humor and was also fond of his morning routines. One such practice involved the dreaded bathroom break on the bay pen following the morning "jo." Grinning ear to ear, Mr. Iceland would arrive to the bay pen exactly on schedule and promptly report to the "Incinolet" or incinerator toilet (very environmentally friendly for the ocean; as for the atmosphere . . . not so much). One need not spend too much time contemplating the inner workings taking place deep within the bowels of the Incinolet.

The bay pen's only "head," the incinerator's vent stack rose from the top of the research shack where another morning routine was taking place: the ethogram, a twice-daily collection of research data. (Later the ethogram became an hourly duty.) The lucky individual charged with taking the morning ethogram recordings would perch atop the research shack gaining a bird's-eye view of Keiko and his activity. He or she was sequestered there, clipboard in hand, until the fifteen-minute observation period expired—no excuses, no change of venue. Consistency is important in data collection. The observer would record numerical values representing various activity levels, time of the observation, any unusual behavior and multiple other raw data points.

Smari's ability to synchronize his morning ritual with the somewhat varying schedule of the ethogram was uncanny. We often theorized that a "mole" was in our midst reporting the day's plan to Smari in advance. Somehow I was always "busy" and skirted the responsibility of recording ethogram data. I was therefore spared the eye-watering cloud bank of smoke coughing out of the vent, assaulting and insulting our beloved researchers engulfed aloft.

Equally endearing, and less likely to prank, was the Icelandic business manager for all on-site operations, one Guðmundur Eyjólfsson. No one called him by anything other than simply "Gummi." Another blond and fair-skinned Icelander, Gummi wore his hair shaved very close to his head which produced a white-halo effect that went nicely with his role. Not a small man, larger than average build and slightly barrel-chested, Gummi looked the part of a very capable and sturdy Icelandic male.

Gummi was in his forties, somewhat older and more mature than the average age onboard the release team. He provided a much-needed keel of competence and seriousness in the otherwise constantly mischievous seas that battered the project atmosphere. Whenever we needed something administrative, had complications with media scheduling or required replacement gear, Gummi was our savior. Gummi was also one of the few that recognized the improvements in Keiko's disposition and believed in the experience that Robin and I brought to the project. His support was unflappable throughout the most trying conflicts within the organization.

Some of my fondest memories of Iceland and the project took place at Gummi's home. Away and apart from our normal surroundings, the contrasting warmth of the quaint Icelandic dwelling was ever inviting. Dark-paneled wooden walls and built-in shelving adorned with collectibles of a northern flair framed each adjoining room from the kitchen to the dining area. A fireplace centered the family room where we sat on an assortment of rocking chairs, a big leather chair, velvet padded armchairs and couches festooned with ornate coverings and piled high with a variety of accenting pillows. Following a meal, Gummi often pulled aside some of the mismatched chairs in the dining room where a few of us savored a cigar and brandy. Gummi and his wife never shied from offering the welcome escape in the heartwarming surroundings of their traditional Icelandic home.

Given the sheer concentration of time together, a few months on the project was all it took to forge lifelong friendships. Ingunn and Siti, security employees for Smari, were frequently stationed on the bay pen for night watch. It took a goodly while for Ingunn to become comfortable around me. I did not know at the time, but she was spoon-fed a certain perception long before she and I ever had the chance to get to know one another. In the ever-changing progression of the release plan, Behavior Team members rotated shifts throughout day and night in order to implement the shaping of Keiko's behavior around the clock. It was during a handful of night shifts together on the pen that Ingunn finally realized I was not the devil incarnate.

Ingunn stood out in a crowd of Icelandic women, and men for that matter. She was taller than average and a redhead, the only person with ginger hair I ever met in Iceland. The quiet type, Ingunn would have been a challenge to befriend were it not for the seclusion on the pen in close quarters with myself and Siti. Raising the degree of difficulty, neither Siti nor Ingunn spoke much English, or they were shy about using the language. I was never sure which. Either way, it was hard to resist the stupefied comedy that has a way of seeping out in the earliest hours of the morning after a long night without sleep.

At first it was Siti and I, looking for things to pass the time, inventing physical challenges that made us both look quite silly. Outwardly, Siti appeared like a rugged old-school father figure. His unassuming quiet nature lent a great deal to this perception. But the more we spent time together, the more I realized his inner goof gave my own a run for its money. It didn't take long to entice

Ingunn's participation in our games. Heck, there wasn't much else of interest competing for her attentions. One night in particular it was unusually cold and damp. We alternated racing around the perimeter of the south pool and timing each other, Ingunn officiated. Somewhere between challenge and boast, Siti ended up running the course in nothing but a pair of whitey tighties. It had to do with a bet. A bet that Ingunn instigated.

Thereafter, we shared some pretty silly exchanges. We often found ourselves in stitches and barely able to breathe for laughing. Never at a loss for being the class clown, I had a great audience in Ingunn. Laughter was the great equalizer, the best medicine for many hours spent isolated on the bay pen, often in weather that kept us in a constant giddy high-adrenaline state.

One of the most reliable people on the project, Siti was not only part of the security detail, he was also an accomplished boat captain and often assisted in the open ocean boat formations piloting one of our support vessels. Siti might have been in his late forties. But it was difficult at best to be sure sometimes with the hard-weathered men of Heimaey, who might easily appear older than their age. We were charmed by his durable sense of integrity and tickled when he often referred to his father as "Daddy."

On the other end of the scale was Hai. Along with his boyishly mussed hair and equally youthful mannerisms he had the energy of a teenager, despite being well into his late thirties. Hai was an exercise in frustration tempered by a healthy dose of dependency. We would have been lost without Hai, but at the same time he was capable of doing something out of left field at any given time. Mostly this trait was a source of levity as Hai was truly harmless, even when he did throw us the occasional curveball. More often than not, it was his enthusiasm of starting new projects long before we had decided they would be needed.

Our Icelandic troops were one of the true highlights of the project; unforgettable characters and experiences that resulted in a lifelong affection for Vestmannaeyjar. Despite many other negatives that haunted us and Keiko during those fateful few years in the Land of Fire and Ice, these relationships would leave a positive memory in their wake that would not be undone.

Target Practice

Though we forged many friendships in Heimaey, there were those that did not want the Keiko Release Project in their hometown and wanted nothing more than to see us, as well as Keiko, go away for good. Sometimes the sentiment translated to bold threats, such as shooting Keiko and turning him into dog food. Most were directed at Keiko, but a few also involved the staff. Initially this concerned us as there were too many vantage points surrounding Klettsvik Bay; hundreds of locations on all sides from which a rifleman could easily pick his target. Our reaction: to increase watch from the bay pen and sometimes the overlook opposite the bay.

Those of us on the bay pen joked about the intelligence behind this strategy. We effectively put ourselves in plain view of any would-be assassin. Stephen Claussen made great fodder with the arrangement, putting a handmade bull's-eye on his chest and acting out the "human target." The threats never turned serious and no one (to my knowledge) was ever caught snooping about the hilltops with malicious intent. In fact, the most dangerous suspects were the mountain lambs that grazed along the sheer cliffs of Klettsvik Bay. The occasional lamb, when it lost footing, plummeted into the bay near the rock face. This was a rare occurrence, but should one get "lambed," it would undoubtedly constitute a life-threatening event. More bark than bite, the threats were nonetheless a telltale barometer of negative sentiments that eventually led to real danger for Keiko.

All things considered, June through August 1999 yielded net positive results. Progress continued with Keiko in terms of his physical exercise and the initial steps of the reintroduction plan we were systematically pushing along.

If any chance remained of capitalizing on the relative calmness and extended daylight of the summer season, it was time for a bold step. It was time to conquer the largest operational challenge facing the project to date. It was time to move Keiko from the small pen to the expansive surroundings of Klettsvik Bay and begin the next phase of reintroduction conditioning. For the first time since Keiko's arrival in Iceland, real progress was about to happen.

All too quickly we found ourselves in the middle of August, days and weeks clicking by at an increasing tempo. Though a concept and design had been agreed upon, much was left to be done before actual work could begin, not least of which included an estimated cost of the enclosure and the board's approval of same. But it quickly became apparent that no matter how hard we all worked or how perfect our efficiency, we were now facing a winter install. More than a few questioned the plausibility, even suggesting that we wait another season. But those of us on-site knew Keiko couldn't afford another year sequestered inside the restrictive bay pen. Somehow we had to find a way.

Overcoming the challenges posed by weather and currents and installing what equated to a giant underwater sail took extensive planning . . . and time. More of the latter than we cared admit. Thus far, the barrier net was no more than a gleam in the eye of the release team. Our excitement and anticipation would have to keep us motivated through the long winter ahead.

6

The Surge

Solarium

At the end of summer we received news that our lease was up with the fire department hostel. Our familiar makeshift home was at end. Thankfully, Robin was the manager at the time, and he hit a home run in landing a rapid solution. No more than a few short blocks away from the hostel sat a newly renovated four-story hotel (actually, still in the throes of final renovation). Despite the ongoing work, our new abode was filled with comforts and accommodations far surpassing the more utilitarian rawness of the hostel. Each staff member got his or her own private room (shared between rotations) complete with private bath. No more locker room showers shared by the same sex. A small but welcome adjustment, even if the "kit" showers of the hotel were, as my father used to say, "so small you couldn't cuss a cat without getting fur in your mouth."

The first level was primarily an entry foyer from the main street. A spiral staircase just inside led to the second, third and fourth floors. The upper floors consisted mostly of individual rooms, although the second floor also had a large kitchen and staff dining area toward the back-street side of the building. On the top floor were two spacious penthouse suites, complete with bathroom, mini-kitchen, bedroom, sitting area, private balconies and an advantageous view of the town. One could even get a small glimpse of the channel leading into the harbor standing on the north-facing balcony, which became a valuable "crow's nest" for assessing conditions in the bay. Given our positions on the release team, Robin and I, and Jeff and Jen shared the two penthouse suites. Typically on opposite rotations, Jeff or Jen occupied the larger of the two while Robin and I split our time in the other. Although I spent the vast majority of my time on the bay pen itself, when schedules allowed, the luxury of the penthouse rooms lent much to maintaining sanity and composure during the more trying times.

The privacy afforded by the rooms also allowed each of us opportunities to escape each other's company, a healthy benefit when working in such close quarters with even the most pro-social compatriots. Other features conducive to group social activity and increasingly frequent meetings provided the perfect balance for a harmonious living environment. The "solarium," a large common area occupying nearly half of the third floor, quickly became our favorite place to congregate. The room's marquee feature—a third of the ceiling and back wall—was comprised of glass panels creating an atrium with unbelievable views of the skyline, complete with surreal northern lights in the dark of winter. Comfortable sitting chairs, sofas and a pool table completed the solarium's creature comforts. This space would be filled with lasting memories, from holiday parties to hard-fought battles over project obstacles. It was second only to the bay pen in providing a backdrop to the ongoing release effort.

The timing of our move to the hotel was, in retrospect, immaculate. Beyond features conducive to the mental health of the team, it also provided momentary distraction to the staff, wearied from the barrage of operational change that had become the norm. Equally as valuable, it provided room for growth.

Battle Lines

On the frontlines in Vestmannaeyjar, the project appeared to be moving in all the right directions. Acceptance of the barrier net plan gained steam. Staff settled into the hotel and their new rotational schedules. Keiko sustained ever-increasing levels of exercise just as his activity levels outside of human directed interaction continued to improve. Were it not for the conflict building with Dr. Lanny Cornell on the subject of the release plan, all would have seemed right for once. But it wasn't to be. Jen and I had outlined the formal release plan, forensically describing each aspect of the reintroduction strategy in writing. The final document was to be submitted for peer review and eventually become the permit submission for formal release approval from the Icelandic Ministry of Fisheries. Disagreement between Lanny, and Robin and I escalated with every detail put to paper.

E-mail Excerpt: August 20, 1999

> *To: Charles Vinick*
>
> *Subject: Re: Lanny*
>
> *From: Robin B. Friday, Sr.*
>
> *Charles, I appreciate your intuitive understanding of our frustrations. Speaking only for myself, I realize Lanny's grasp within the board structure. Jeff and I are constantly throwing ideas at each other with respect to this scenario. Should we gather for a presentation to the board, no matter what Lanny may say to you privately, he will unload in an open forum. He has put his feelings in front of them before and he is not an individual to defer to "incompetence."*
>
> *My impression is Lanny is making a stand. Why? I wish I knew. Charles, anyone, from a first year apprentice, would consider his theory totally irresponsible and at the brink of blatantly violating his professional ethical standards.*
>
> *It appears most skeptics are pointing at the behavioral approach, yes, training. Why, they say?*
>
> *Because that is what experts around the world are saying should be done.*
>
> *No one is asking why we are not consistently documenting his physical well-being. It wasn't behavior modification that almost killed him, it was*

> *health concerns. If a human patient had been in critical care within the last twelve months, under the treatment of the strongest antibiotics available, had a history of chronic papaloma viral condition, and was being prepared to make it on their own, one final treatment, then no more doctor, do you suppose this is the level of care that individual would receive before the door was shut off?*
>
> *Thought I was going to stop complaining, so did I, sorry. We will brainstorm more ideas. I will forward Lanny all recent developments of marine logistics and associated information.*
>
> *Hope you're smiling, Robin*

Between the raging battles playing out on the computer screen and the never-ending meetings on the finer points of the barrier net, Keiko himself began to exhibit troubling signs. After months of progress and reaching new levels in physical stamina, he suddenly slowed. His interest in everything had dwindled, including food and the normally intriguing activity going on around the bay pen. His diet went from 120 pounds of fish a day to less than twenty. We saw a return of traditional lazy behavior such as floating at the surface and slow responses during interactions with the Behavior Team.

Going off his food was not a good sign. It could be related to almost anything, but often the first concern is physical well-being. Was he sick? Was Keiko's system fighting something? On the heels of recent foul weather and endless changes implemented in his conditioning program, it would not be uncommon for stress to compromise Keiko's immune system. That, added to the surge currents stirring the pot, quite possibly introducing a plethora of bacteria or pathogens to Klettsvik Bay, meant anything and everything had to be considered. In August we had seen a flare-up of Keiko's skin condition, a papillomavirus that acted much like the human herpes virus. Although the condition did not advance to the ugly cauliflower-like growths they had witnessed in Mexico and Newport, the early-stage pinholes appearing on certain areas of his body were surely indicative of a weakening immune system or heightened stress levels.

Normal procedure dictates the collection of various diagnostic samples on a routine basis, such as blood samples, blowhole cultures (like a throat culture), cytology, fecal and urine samples and a host of other clinical metrics. Like many zoological animals, Keiko was well trained to provide these samples voluntarily. In the professional field of animal care, these sometimes-daily routines are called "animal husbandry." Husbandry diagnostics are the front line in preventative care. Since animals cannot tell us, "Hey, my tummy hurts," there are few other means by which to proactively discover and treat a potentially serious health threat. Even the most innocuous event can become life threatening if undiscovered and untreated long enough.

Yet no routine medical evaluation existed for Keiko. Largely because he was considered a temporary resident, Dr. Cornell played a dangerous game of "don't ask, don't tell" with Keiko's health. He did not pursue any regular husbandry schedule. In the time I spent with Keiko, blood samples and other health measures were only taken when there was a preexisting concern . . . a red flag.

By the time Keiko had lost interest in his food, it was time to be very concerned. Why? Waiting until an animal loses interest in eating to investigate is like waiting until steam is coming from under a car's hood only to find out that the engine is overheated. Many times it's too late to reverse the damage. As gauges are to an automobile, so too is husbandry to zoological care. Constituting an end-run around Lanny, Robin immediately took blood samples from Keiko and had the basic levels analyzed at a hospital in Reykjavik. This action alone flew in the face of protocol. Yet Robin knew any samples sent to Lanny would be ignored or worse, reported as "within normal limits." It didn't take long before Robin had the results. We were lucky . . . this time.

Keiko's white blood cell count was normal, and his blood showed only a very slight decline in hydration level. The normal white blood cell count told us that he was not fighting an infection, and the mild deviation in hydration markers seemed to be only in response to his recent drop in eating. Normal for a killer whale, Keiko's only material source of water was the moisture in his fish. If he didn't eat enough, he would soon become dehydrated, starting an avalanche of other medical problems. Maintaining good hydration is paramount in proactive health maintenance for any animal or person.

Our initial fears were allayed, but the problem was far from solved. What had caused the appreciable change in Keiko's behavior and his loss of appetite? In zoological care, there are standards for evaluating this type of mystery; a process of elimination, analyzing the usual suspects and going down the list.

Among the battery of customary tests, we ran water quality sampling, cultures around every conceivable husbandry area and finally analysis of Keiko's food.

It did not take long to find the culprit. After receiving the nutritional analysis results, we realized that a the original herring lot had been swapped with an untested lot. The freezer-house manager didn't even consider that the change in fish mattered. After all, what could the difference between one box of fish and another matter to a whale. In fact it made all the difference! The new lot replaced the former 440 kilocalorie-per-pound fish with herring much richer in fat and nearly double in caloric content. Unbeknownst to us, the key ingredient in Keiko's diet had changed substantially.

Each lot of fish is different. Two differing lots may be caught at opposing times of the year and in different locations. They may even be processed and frozen using varied methods. All of these discrepancies have an impact on the nutritional content of the fish, from protein and ash to water content, and of course, fat and calories. Somewhere in the recent past, likely weeks prior, the lot had changed. We had been feeding Keiko the equivalent of a six-course holiday dinner day-in and day-out. It was no wonder he had lost interest in food. Worse, as he became satiated with the high fat levels, he had also become inactive, the majority of his energy drained in the process of burning off the excess nourishment. We weren't unfamiliar with the symptoms.

Frustration was an understatement. The oversight in Keiko's diet was a stupid mistake and one that would cost us dearly in the advancement of his rehabilitation. Further complicating the situation, we had no other source of herring with which to replace the calorific lot. Our only option was to decrease Keiko's daily intake to less than twenty percent of what it had been. At this level, even if he showed interest in exercise and activity, we had to be careful not to overwork him. The severely reduced diet meant he would not be receiving the water a whale of his size required. We immediately began sourcing a new lot of herring, one low in fat and calories and high in moisture. Ironically, the source we found would have to be shipped from Boston, and the process of importing a fish back to Iceland was not a simple one.

Pressing forward, many of Keiko's sessions involved no food at all. Initially, any interest he displayed in his trainers quickly dissipated if we so much as showed him a food bucket. On one side of the equation, we worked to carefully increase his calorie burn, to break him out of the lethargy, and to help his body to rid itself of the amassing blubber. On the flip side, lacking any better source of food for the time being, we continued feeding Keiko the fat pills. This 10,000-pound animal, typically consuming over 150 pounds of fish a day, was now only able to stomach twenty pounds in the same time. We were in a deep hole, and it would be a long slow climb out.

People's Feast

At least some forms of progress offered a small but welcome respite from the setbacks with Keiko. August saw the advancement of barrier net plans, as they emerged from paper to reality. Robin, Charles and Jean-Michel Cousteau (a famous son of Jacques Cousteau and the head of Ocean Futures Society) had met with net makers in Reykjavik, one of the few possible vendors that would consider the task. Like any undertaking that pushed the envelope on accomplishing the impossible, there were many contractors who would not touch the barrier net for fear of backlash and the risk of liability. Still, the organization was serious enough that the dredge of persistent obstacles had not yet deterred the way forward.

No matter the commitment, progress was agonizingly slow. So many surveys of the bay, core samples, test materials and cross-examination of willing contractors dragged on, stifled at times by a season rife with holidays, traditional celebrations and festivals. One in particular virtually shut down the island of Heimaey for more than ten days. That was none other than the world famous People's Feast.

The celebration began in 1874 on the mainland commemorating the 1,000th anniversary of Iceland's semi-independence from Denmark. Unable to attend because of inclement weather, the people of Vestmannaeyjar held their own small celebration. Eventually, the gala on Heimaey grew vastly more popular than the mainland festival with more than 10,000 in attendance in recent years. I had heard bizarre stories of this festival from staff and Icelanders involved in the Keiko project. I took most of the stories with a grain of salt, as many of the tales surrounding the People's Feast sounded far too extraordinary to be literal. The thousands who attend the feast in Heimaey camp on the island's only golf course (the second northernmost golf course in the world) and create a tent city, a mini suburb, erected overnight for the three primary days of merriment.

Camping, barbecues, bonfires, fireworks, singing, dancing and even festival arts are there for the taking. Not on the published schedule, but informally known as a highlight festivity: sex in the open fields surrounding the tent suburbia. This I had to see to believe. Really. Surely this was but another grand exaggeration.

Not so. In fact, should an Icelander ever scream "duck," do not hesitate to do so, quickly. In my experience, their culture doesn't know how to exaggerate. Most folklore I had heard initially came across as farcical. After all, I wasn't just some gullible American. Wrong. Icelanders don't do April Fools. They don't have any interest in impressing their audience via hyperbole. A 100-mph wind is a moderate breeze. In their matter of fact way, they simply state what is. After witnessing multiple entangled silhouettes rhythmically dancing on the hillside at the People's Feast, I never again underestimated the colorful stories I was told about Iceland. In most cases, it was safest to overestimate.

The Storm

> *Tue 9/9/1999 2:30 PM*
> *To: Mark*
> *From: Robin*
> *Mark*
> *Will call Lanny as soon as this storm is gone. It's going to be a bad one here, my friend!*
> *Robin*

New digs, conceptions of a barrier net, staffing changes and advancing outlines of a formal release plan filled a busy August, brought to a close by the unforgettable People's Feast. The end of summer also meant the arrival of winter's foul weather. On the southern border of the mainland, we were the bull's-eye for all northward bound storms following the Gulf Stream. Wind and rain, the most obvious of the elements that stirred up Klettsvik's innards were not our greatest foes. Within the seclusion of the bay, currents posed the biggest challenge. Any notion of avoiding them was delusional.

The surge currents of Klettsvik were elusive and mysterious. They could occur in concert with a clear and present storm threat, or they could rush into the bay on a bright and sunny day completely unannounced. The frightening strength bestowed an ominous reputation of the surge among the staff. No one took it lightly. No one wanted to be stuck on the bay pen when the nightmarish current struck.

Somewhat in jest, but all too necessary, we often spent time lying on the floor with an upside-down perspective looking at the layout of the research shack's interior. The exercise was intended to give us a familiarity with our surroundings should we ever have need of escaping an overturned and underwater habitat. The possibility of the research shack taking an inverted plunge was not far from reality. Should the bay pen's superstructure give way, the resulting imbalance of the top-heavy segment holding the research shack would cause it to immediately flip. Tangled in anchor lines and tossed by a current strong enough to pull the bay pen to pieces, no one on the team had escaped the thought of being trapped inside a watery coffin. In the less than forty degree Fahrenheit water, even a survival suit can only sustain a person for so long. Certainly not long enough to get a rescue boat to the bay pen in severe weather.

The Icelandic Coast Guard (ICG) contingent in Heimaey possessed one of the most remarkable spectacles of oceanic rescue any of us had ever laid eyes on. A thing of sheer utilitarian beauty, the jet-driven rescue vessel, aptly named *Thor*, was everything anyone on the receiving end of its hospitality could ever hope for, and it certainly lived up to its legendary Nordic namesake. Bright orange, so as to be easily seen under any conditions, the *Thor* was the quintessential masculine mechanical beast: an M-1 Abrams of the sea.

Roughly thirty feet in length, the *Thor* was proportionally thin for its span. Its hull design gave it an appearance of a thoroughbred dancing at the start gate and ready to go all out. The perimeter of the boat was made of buoyant foam-filled pontoons called sponsons. The pilothouse (which consumed a majority of the deck space) sat just aft of amidships, centered on the beam. Upside down, the *Thor* was a submarine of sorts. The pilothouse was designed to be completely watertight. However, if ever capsized the *Thor* would not stay that way long; the vessel had a self-righting ability, meaning it could flip itself upright.

We took the *Thor* out to sea a few times, though never under our own control; the ICG always manned this vessel—no one else. Those few outings were almost exclusively for whale watching or tracking the seasonal movements of the wild killer whales around Heimaey.

Inside the pilothouse, the fixtures and accoutrements clearly illustrated its business-only temperament. Every seat in the cabin was adorned with safety restraints more overzealous than the most extreme roller coaster. Designed to give her pilot complete stability in relentless seas, the captain's chair was afloat on a heavy piston that would mitigate backbreaking jolts common in high seas and foul weather. Every steerage the pilot needed was attached to this shock-absorbing seat, even the singular do-everything joystick control. So stout and convincing was the *Thor* (and the ICG) that it gave me a dangerous confidence at sea . . . as if my only interest in plus-twenty-foot seas was a morbid curiosity of how the *Thor* would power through or dance over the pounding waves.

Icelandic people are among the most proud people I had ever met then or since. But in relation to the *Thor*, pride was an understatement. The ICG's slogan is "Always Prepared," and they are. Few vestiges of island civilization know the challenges of extreme oceanic emergency and rescue as well as the Icelandic Coast Guard. Though I had no naturalized right to be, I was extremely proud of the ICG and its most impressive *Thor*.

Our chief of Marine Operations, and Smari were equally inspiring of confidence. Michael Parks was licensed to pilot 300-ton ships and he possessed a natural aptitude for anything waterborne. Smari, a commercial diver and member of special recovery/salvage teams, had participated in deep-water dives around Iceland that raised the hair on the back of my neck whenever he talked about them. On one occasion during my time in Iceland, Smari had to recover bodies from a shipwreck off the mainland shore at more than eighty meters deep. I could scarcely imagine diving at that depth, in the complete dark, and along with the psychological burden of recovering bodies no less. In the event that we needed rescue from a collapsing bay pen, we were blessed with many reliable resources that could come to our aid.

September 9, 1999, I was at home in Orlando, one of my brief off-site rotations. As usual I received daily e-mails and the occasional phone call from Robin. We routinely dissected everything from

Keiko to the barrier net and beyond. On this particular day and in Robin's classic way of downplaying just about everything, he ended an e-mail with, "It's going to be a bad one here, my friend!" I had known about the storm because Klettsvik was always last in line to receive every storm cell that traveled the eastern seaboard, even those that originated as far south as Florida. This was due to the "Gulf Stream Express." The Gulf Stream, one of the most well-known currents in the Atlantic, carries most storm systems straight up the east coast of the Americas bouncing off Newfoundland and smack into Iceland. Vestmannaeyjar is like the keeper in a soccer (or football) game. The small island chain takes the brunt of every Gulf Stream kick aimed at the goal.

There were two distinct weapons of each significant storm cell that reached Vestmannaeyjar: wind and current. Almost always, the wind came first, the surge current arriving shortly after. In some cases the wind stayed long enough to join forces with the surge current, and both wreaked havoc on the bay pen and our nerves. This particular storm was the offspring of at least two hurricanes, Floyd and Gert, which were lurking throughout the lower reaches of the Gulf Stream vacillating between tropical depression and hurricane status. I don't know that either of these two systems ever actually reached Iceland intact, but there was no doubt that they had fueled activity that reached as far north as Klettsvik Bay on September 10, 1999.

On that day, the resulting surge currents levied a heavy toll on the bay pen, breaking the superstructure in two places on the east side of the southernmost pool. So strong were the currents tearing at the bay pen that the enormous concrete rings that weighted the north and south pools were lifted to the surface throughout the night like giant lids hinged on the sea floor—a sight that was wholly unsettling.

Keiko himself seemed the least affected in such conditions. Often, he would merely float as best he could in the calmest corner of the pen, on the leeward side of the wind or current. In the worst of conditions, he was forced to swim continuously, the only means by which he could avoid being tossed into the moving parts of the pen's structure.

Another time, the bay was blanketed in the darkness of winter. A surge had hit Klettsvik Bay sans any accompanying wind and lifted the several ton rings to the surface, first in one direction and then the other as the current bounced back out the way it had come.

Strangely, the current by itself without the wind was somehow more frightening than the combined duo. I was on the pen during two such occasions; watching forces of nature so immensely powerful, yet so indifferent, that one instantly senses his own frailty. Realization strikes with sobering clarity that at any moment, we could be fighting for our very lives. We theorized that these forceful and mysterious currents were generated by sea floor tremors, otherwise undetectable from land. At times, even on clear weather days, we had watched as boats in the harbor rose up more than ten feet right before our eyes and in less than a few minutes.

Not this time.

This time the surge was the offspring of active storm cells in the North Atlantic, and the devastation levied on the pen was accompanied by hurricane force winds and rain. It was on the eve of Smari and Lina's wedding that Robin's storm warning took effect, twisting and contorting the bay pen to such extremes that two primary infrastructure joints in the thirty-inch HDPE pipe gave way, compromising the pen's buoyancy and shattering the equilibrium of the anchor system. What had served to pull equally on opposing points of the pen, thus stabilizing the structure, now gave purpose to pulling it apart. No one wanted to tell Smari, knowing he would replace his marriage responsibilities with that of rescuing the pen. Discreetly and in haste, Michael led the Marine Ops crew in patching the stationary craft together. Remarkably, they did so in time for the wedding; however, only by amputating the entire southern pool. The damaged end became a liability threatening the only remaining vestige of the habitat surrounding Keiko.

Significant restructuring of the anchor system would be needed in the following weeks. The damage and subsequent repairs forced us to shorten the perimeter itself and close off the southern end preventing Keiko's access to that area of the pen. It would be some time before the south pool could be reopened to Keiko. Ironically, in the approximations toward life in the open ocean, Keiko's world had just been reduced by half. For the time being, Keiko would be sequestered to the north pool, adding additional challenge to increasing his activity and physical workouts.

The Gudrun

One lazy afternoon, I set out on foot about the town to clear my mind. A rare escape. There was no purpose in my direction, only that of solitary and pensive meandering. Not long into my brief respite, I happened into the far southwest side of the harbor, where the big ships were moored. These were impressively large working-class ships. I was enthralled with the lifestyle. As I often did when surrounded by seafaring vessels, I would stand, staring at the ships with their rusty outlines, hard steel covered with countless layers of paint, willing myself to relive colorful scenes played out upon their decks. They were scarred prize fighters—brutes, hardened by a life struggling against a sea obsessed with swallowing them whole. I imagined what it must be like: what the lifestyle would feel like, and how the fisherman that occupied these leviathans for weeks and sometimes months on end might view the world. It was not an easy way to make a living, but not hard to imagine that it also created a very strong brotherhood among fellow fishermen.

Walking around the very end of the harbor on my way to leave, I saw her and stopped in my tracks. The *Gudrun*. Instantly I had goose bumps, the kind that sent a wave through my entire body. I didn't know what to think. A mixture of feelings erupted and coursed through me as I tried to form a cohesive thought. This was a piece of history I never expected to see firsthand. The *very* ship that had been used to collect nearly all the killer whales from Icelandic waters in the 1980s and early '90s . . . and there she sat, just like that. This was the boat that escorted many of the whales I had known on their inaugural transition to a life with humans. The

Gudrun was definitely where Keiko started his journey. I stood, staring for an unknown length of time, imagining and wondering, trying to resurrect the scene and immersing myself in the significance and enormity of it all. The vision of the *Gudrun* was surreal. It was a relic of history specific to my life, to my career and to the many amazing relationships that I had been blessed with over my time working alongside killer whales.

This inanimate salted beast of steel was in many ways the centerpiece of one of the most remarkable animal movements in human history. She had ferried animals from the waters of Iceland and thus started them on individual journeys that led to various corners of the world. People of all walks of life came to know these animals from Iceland. The care, compassion and intrigue they excited led many more to the spectacle of *Orcinus orca* and still others to a life of protesting their confinement. In any case, the *Gudrun* was an unsuspecting player in the rise of value for killer whales worldwide. I wondered if her captain had any clue how far the activities of his battered vessel traveled or how they have altered the values of entire societies and generations of people.

From Florida to Iceland

September 1999 we introduced a new hire to the Behavior Team. Kelly Reed had come directly from SeaWorld and a career of working with killer whales in an applied behavior setting. Hiring Kelly was the product of much deliberation between Alyssa and me, and only after considering almost a half dozen others for the task.

Alyssa was gifted in her ability to assess true talent in our shared professional field, and I trusted her faith in Kelly's expertise. Although I had never met Kelly, I spoke with her at length, often deliberately trying to dissuade her interest in the project. I knew one thing from firsthand experience: this job was not for the faint of heart nor the thin-skinned.

Of particular interest to me was Kelly's tutelage under Ted Turner. Not the Ted Turner of cable news, but the Ted Turner, vice president of Animal Training of SeaWorld of Ohio. In the comparably smaller community of zoological professionals, Ted's fame was nonetheless equal to that of the Turner Broadcasting version. Many of the most complex behavioral achievements in the SeaWorld system were the product of Ted Turner's obsession with the science of behavior and the application of behavioral modification techniques. He was famous for the intensity with which he imparted the profession on his pupils. I knew that Kelly would be an asset to the staff in Iceland, capable of not only understanding the behavioral rehabilitation already in progress with Keiko, but also providing additional experienced input. An added bonus, Kelly had just weeks before been caring for three other male killer whales in the SeaWorld of Florida park. This abrupt juxtaposition would produce a "State of the Union," a very useful contrast in evaluating Keiko's behavioral and physical disposition.

In her early thirties, Kelly was a lively addition to the staff. Full of excitement borne of the project's novelty and exotic climate, she made everything new again. Kelly brought a fresh perspective, and her excitement was contagious. Admittedly, I was so long buried in the sensitivity of the existing staff and the pendulum swing of morale, that I welcomed Kelly's arrival more than I could admit. At last I had someone who spoke the same professional language, understood behavioral science and, as importantly, recognized that it was a priority in the management of any animal environment, most especially this one. When I introduced a thought, alternative, or solution regarding Keiko's daily interactions, I didn't need to preface it with "why." The freedom to openly discuss behavioral modification without accusation of speaking "SeaWorld" or risk of alienating my skeptical coworker was liberating.

I don't think Kelly weighed more than a 105 pounds. She was an attractive blonde with a thin figure appropriate for anchoring a model runway, but she didn't have nearly enough mass to safely navigate the winds of Klettsvik Bay. Case in point, I weighed in at over 200 pounds. On at least one occasion with two thirty-five-pound fish buckets in hand, I had been completely lifted off the bay pen deck by wind gusts ricocheting off the sheer rock walls surrounding the bay. Kelly wouldn't stand a chance if caught unaware.

Kelly hired onto the project at an unusual time. Everything was in a state of constant change. From the release of a favored team member and the segregation of duties into specialized teams to the modified rotation schedule and the ramp up in severe weather operations, she could not have started at a more challenging moment (for the existing staff). It would do no good for Kelly to shadow me on the bay pen. Very quickly I needed her to get up to speed and begin carrying responsibilities in my absence. Her first foray in Heimaey would be the only rotation where we would work together for such an extended period. Thereafter, I placed Kelly on an opposite rotation and paired her time on-site with Robin. Her presence would allow Robin to focus his attention on the Marine Operations team and achieving the impossible: the installation of the barrier net.

To my surprise, fortune favored our timing. In earlier months Kelly's addition to the team would have resulted in a family feud the likes of which no person should endure. However, in the midst of the energy and enthusiasm surrounding the prospect of bay access, Kelly's start produced no more than mild scandal. Still, the contrasting personalities of Kelly and Steve Sinelli were nonetheless quite humorous. Like baking soda and vinegar, the two never failed to elicit a spirited reaction from one another.

Kelly possessed a rather unusual brand of humor. The uniqueness came from the fact that most of her pratfall-type comedy was unintentional. Kelly's propensity for physical mishap was at its peak in an environment like the Keiko Release Project where opportunities abounded for happenstance goof ups. From dropping the heavy stainless steel buckets into the harbor to stalling the transport boat right in front of the oncoming ferry, Kelly was always giving us raw material from which we gained much levity.

Fortunately Kelly's misadventures were never dangerous, but they were colorful and varied. Some of the more comical moments in our tour of Iceland were at her expense. In its own strange way, this hallmark trait of Kelly's eventually endeared her to many of the staff. No matter, Kelly was not hired to manhandle the obstacle course of the physical environment; she was hired to bring much needed balance and expertise to Keiko's behavioral rehabilitation, and that she did.

Release 101

September came and went, accelerated by the activity of the bay pen reconstruction and the novelty of a new team member. At the same time, issues surrounding the formal reintroduction protocols came to a head. This was an important milestone in the project not to be underestimated. The document would constitute the first time the release plan would be published and reviewed by peers within the scientific community. It was a point of no return for the organization. By putting the plan in writing and sharing it with professional colleagues, FWKF was committing to specific methods and metrics for evaluating the viability of release.

Perhaps more impactful, this document would become the application for final approval from the Icelandic Ministry of Fisheries for Keiko's introduction to the ocean and eventually wild whales. U.S. Fisheries also maintained a complimentary right of input on the documented plan. The protocol document would have to detail what constituted a successful release and define an intervention response plan in the event that Keiko failed to thrive after release. Divergent positions from within project management on every aspect of the process were steadily rising to the surface, driven by the need to put consensus in black and white print. In every case, the most vocal opponent to our plan (already well-under way) was Dr. Cornell.

Point by point throughout the draft document, Lanny attacked the plan. Never one to lower himself to the actual work of correspondence, Lanny extorted his opinions through Charles, who in turn relayed the most salient points of the doctor's judgments through e-mails that lobbed back and forth across the continents day and night.

Charles' first volley from the High Court of Cornell denounced any and all behavioral modification objectives and insisted that we "avoid tying the behavior modification to very specific, clear objectives." We were appalled. In the absence of defining clear objectives, how then could Keiko's readiness for release be measured? Further, Keiko's reintroduction had been veiled in the guise of a scientific undertaking, one of great pomp and circumstance under diverse scrutiny. The scientific method, in its most rudimentary form, requires the ability of an entire experiment to be reproduced independent of the original study. In his customary straightforward way, Robin responded.

E-mail Excerpt

> *September 29, 1999*
> *This project is not an everyday event. The behavioral terms are described in basic and generally accepted psychology terms. In my humble opinion we cannot over simplify a complex process any more than a rocket scientist describes a trip to the moon as "putting gas in a can and lighting it."*

Ultimately, the logistical and behavioral plan of approach would be mandated via the official Keiko Reintroduction Protocol document, but it was not without a fight and constant fringe battles. Lanny not only directly disagreed with most of the document, but orchestrated a showdown at the FWKF board level regarding many of the issues. By all outward indications, Lanny simply wanted to "cut the net." The arduous process of planning and sharing that plan with industry and professional colleagues appeared to be beneath him. He had no patience for the process, and it showed in his comments. Moreover, he took every opportunity to "simplify" the proffered plan. If the organization would be forced into the responsibility of a formal plan, Lanny made every effort to ensure that plan said nothing that could be criticized or overanalyzed by people in the know.

In the annals of the marine mammal field, Lanny could undoubtedly stake claim to more than a few advancements in the management of zoological animals, especially killer whales. But the idea that he had any conscience for Keiko's plight was a fallacy. His disposition toward the animals in his care repeatedly pointed to a man driven more by ego and personal advancement than any altruistic intent.

In the folklore of SeaWorld, his alma mater and the setting of his ascension to notoriety as a vet, one such story exemplified Lanny's predilection, some would say, of self-preservation over any ethical responsibility toward animal welfare. Following an animal death in the park, the veterinary team reporting to Lanny completed a necropsy investigation on the animal's carcass, required by law and intended to identify the cause of death. In the gross morphology, nothing overt was found. The next phase of the investigation focused on pathology, microbiology and other microscopic means to identify the unseen culprit. Lacking visible evidence, the final analysis would likely point to something bacteriological or viral that had been in the environment.

Lanny joined the investigation only after the veterinary staff completed a thorough dissection. Saying nothing, he reached into the midsection of the carcass and felt around the animal's adrenal gland. A few moments later he pulled a rifle bullet from the body. Then and there he declared the cause of death an abscess infection resulting from a fisherman's vengeance suffered long before the animal belonged to SeaWorld. The vet staff was stunned. Driven by the moral responsibility of their chosen profession first and secondly by fear of Lanny's hatred for incompetence, they had combed through the corpse in painstaking detail. Yet in the hours they spent searching, nothing of the sort was discovered. Sharing a common thread among the tales that have coursed through the inner circle regarding Lanny's fabled past, the *convenient* discovery effectively averted any personal responsibility for the animal's death.

Now responsible for Keiko's release, it was just as likely that Lanny would make sure no one had a noose with which to hang him as the head vet on the project. If he could stop the advancing plan, it would be at the FWKF board level, before it reached outside colleagues in the peer review process.

We struggled continuously to educate the managing members of the organization. The board had Lanny in one ear—selling Keiko's release as overly simplistic—and us (through Charles) in the other ear presenting exactly the opposite, in agonizing detail. I believe that Charles recognized the sense in our approach and thus valiantly represented what he understood to be a responsible medium to Lanny and the board. But Charles could only regurgitate so much on his own. In spite of his communicative talent, it was not his area of expertise. Eventually it became evident that Robin would have to meet with the board himself and address the most stubborn issues that continued to impede progress.

Pressing on in his dissection of the formal release plan, Lanny further criticized the plan's intent to improve Keiko's physical stamina. Moreover, he summarily dismissed the idea as one of object threat. In his argument, afforded by his experience in similar situations, he claimed that Keiko could "shut down" if too much effort was required. Additionally, and in stark contrast to the stated goals of release, he wanted any reference of complexity in Keiko's transition to eating live fish removed from the document. This time Robin took great pleasure in addressing the salient point on "too much too fast."

E-mail Excerpt

> *September 26, 1999*
>
> *Lanny is absolutely correct in his statement that animals can shut down if too much effort is required for food, or anything else for that matter. It is called "abulia" and can also be described as "learned helplessness." If the task is too large, or the change too great . . . the animal experiences a loss of will power and "shuts down" sometimes refusing to eat, respond to normal environmental stimuli, or even move. This effect has been noted and studied at length in humans and animals. Interestingly, this concept is the foundation of the*

> *argument behind an incremental reintroduction effort as opposed to the "cut the net" approach. Life is fun isn't it . . . if you stick around long enough someone will put their foot in their mouth and not even realize it.*
>
> *There are statements within this document that oversimplify this process. I fear that these statements to the board and the media can and have misconstrued the complexity and level of difficulty involved in the reintroduction of a captive killer whale to the wild. We have shot ourselves in the foot and done a disservice to the public to speak of Keiko as being "ready." FWKF jumped the gun on announcing that Keiko was eating live fish and by miscommunicating the importance of live fish consumption when at the moment we are just trying to get Keiko to eat period.*

Comfortably oblivious to the tyrannical exchange with Lanny, the Marine Operations team continued in its own world moving toward the installation of the barrier net. Michael Parks, our chief of Marine Operations heading up this process, was obsessive-compulsive about safety. When working in a marine environment around boats, lines and the unpredictability of the sea, a healthy respect for Mother Ocean can be of great value in keeping all appendages intact. The last thing the project needed was a serious injury or, God forbid, a death.

Everything about the barrier net was an exercise in overcoming obstacles. Nothing came easy, and this was the prevailing challenge even before installation of the net began. The first hurdle came in determining what material could withstand the forces that would wreak havoc on the net itself. Woods Hole engineers had calculated that the net would be subjected to force currents exceeding eighteen knots. Deciding what the net should be made of was a critical decision. It would need to be stronger than steel, but flexible and light.

Another important decision involved how large the "netting" had to be . . . in other words, what size to make the "boxes" created by the crisscrossing of the net material. The smaller the net's mesh, the more material and thus more drag on the net from currents in the bay. Too large and the net would not be strong enough to hold together. Another incalculable factor was natural and artificial debris. Any seaweed, kelp or trash that collected on the net would only increase its drag coefficient and put additional stress on the net and anchoring system. This had as much to do with maintaining a clean net, but in the design it was also a factor in determining the size of the mesh.

If successful in getting the barrier net installed, it was abundantly clear that it would become a maintenance nightmare to keep in place. Soon, our heroic Marine Ops team, aptly nicknamed "Mighty Mo," would redouble their efforts to maintain the tools of reintroduction. They were well acquainted with a bay pen always on the verge of ripping apart. Now they were inheriting an 800-foot monster net with an unknown appetite competing for similar affections.

By the end of September, Michael, Robin and the Marine Ops team were working with OFS marine technicians, Vestmannaeyjar net makers, and several local and international experts in the art of marine construction to answer the unknowns. How could the net be anchored, but allow for the tidal variance? What would be used to actually anchor the net in the rock on each side of the bay? How were we to get boats across the net? How would the net be visible to other boats? When ready, how was Keiko going to exit through the barrier net?

In the following weeks, each question would be answered or at least theories would prevail. Many of the solutions could not be tested until the net was physically in place across the expanse. In all cases, the true test of the net would come under the often extreme conditions particular to Klettsvik Bay. Designing and installing a net of this size was not such a monumental task when the seas are calm, but in the formidable surge currents that plagued Klettsvik, no one could truly know if their theories would hold water until after the fact.

At Long Last

It was the eve of October and we had been on an uphill battle with Keiko's diet for six long weeks. While food alone is not a compelling motivator, the excess of food (satiation) can be an overwhelming force in deterring motivation. Without question we had been at a literal standstill in Keiko's physical rehabilitation. Worse, we were still feeding the rich, high-calorie herring, still awaiting the arrival of a viable replacement.

During this satiated period, Keiko had no interest in food; therefore our only means to alter his behavior came through other forms of reinforcement we could provide. He had become so pigged-out on the "chocolate cake" herring that his old and lazy pastime of boat watching rose to an all-time high. Still hopelessly locked within the confines of the bay pen, our menu of reinforcement was narrowed to no more than our relationship with the Big Man. We had to move him more to mitigate the fattening barrage of calories that had nearly reversed the summer's progress. But as the reality of the barrier net came slowly into focus, this need for increased human interaction was frustrating. It seemed the whole of August and September had been lost to stagnation and Keiko's backwards interest in boat traffic.

E-mail Excerpt:

> *October 3, 1999*
> *Subject: Our boy Keiko*
> *To: Brian O'Neill*
> *From: Mark*
> *Hey, Brian,*
>
> *I think we may be on the brink of breaking through to the other side with the big man's hunger drive . . . going to take him to 44 today and possibly hold there. Problem is that we are still feeding the fatty food. Should have capelin to substitute by today. Robin has located good herring in Boston and we are trying to ship it in ASAP. That will fix one of*

> *our problems . . . the rest are just time.*
> *Take care . . . enjoy that beautiful state of yours . . .*
> *Mark*

As fortune would have it, October would see a turn in our favor on several fronts. Our noble athlete, forced into dragging lethargy, finally seemed to be emerging from his slumber. At the same time, we received word from the west that other hurdles had been overcome. Namely, Lanny's stand at the "OK Corral" board meeting had run its course. The general outline of the release plan had been ratified. Soon enough, with the doldrums of summer behind us, looming changes imposed by the approved release plan would begin to electrify the Land of Fire and Ice.

Paralleling the foundational work of preparing for barrier net installation, the Behavior Team was hard at work on our own small pet project: a means to give Keiko access to the medical pool. As a part of the repairs made to the bay pen in early September, the medical pool had been completely netted off in order to keep Keiko away from jagged and harmful areas of the pen damaged in the storm. A general truism of life, problems often come hand-in-hand with opportunities. By placing a net wall on both sides of the medical pool, the Marine Ops team had in effect provided us the means to build a gateway into the medical pool. The configuration would allow us the chance to train Keiko on going through a net gate in anticipation of the same access method to the bay beyond.

This was an important step to rehearse. Keiko had a history of problems with going through gated channels. In his Newport facility, the trainers had much difficulty getting him to go into that facility's medical pool. The Newport med pool had a gate on each end, and Keiko would swim into the med pool with his entire body, leaving only his tail flukes just outside the gate through which he entered. Rather than swim through the medical pool and out the other side, he would slowly back himself out of the same gate. This behavior was the trademark result of a traumatic history with gates somewhere in Keiko's past. During his brief habitation at

Marineland in Canada, other whales, his dominant pool-mates, repeatedly bullied him by pinning him behind gates and in smaller pools. He learned that by leaving his flukes in the gate channel, he could physically block his caregivers from being able to shut the gate, thus never committing to going into the smaller medical pool.

In my time at SeaWorld, conditioning and maintaining gating and separation behavior was a daily theme, incorporated into every session and interaction. Teaching an animal to hold for gates to open and close or voluntarily move between pools and/or social groups was made to be one of the most fun activities in which the whales engaged. The fact is that proper gate conditioning, as simple a concept as it seems, eludes many caregivers who work in managed care environments. All too often coercion and baiting become the well-worn tools of separating animals. A product of this shortcoming, Keiko's history with gates, mainly avoidance, was typical of the behavior produced by many smaller old-world facilities.

Why was this avoidance of gating an important obstacle to overcome? In this case, it was vital that Keiko commit to entering the med pool in order to fit him for the tracking device prior to open ocean access. Even more relevant was that Keiko learned to go completely through a narrow gateway. In order to get him bodily from the bay pen to the wide expanse of the bay, and then again from the bay to the open ocean, Keiko needed to be fluid and proficient at going through gates.

A side benefit of conditioning the gateway was the mental stimulation this new goal would provide Keiko. After only a few weeks stuck in the north pool alone, amplified by his recent dietary setbacks, Keiko's day had become painfully monotonous. He was beginning to show signs of withdrawal. We couldn't afford a setback of this nature . . . not at this stage in the rehabilitation. The timing of the med pool gate was serendipitous, a perfect and welcome change of pace for the Behavior Team and Keiko alike.

Excerpt
OFS Public Web Update: November 2, 1999

Behavior modification was first applied to cetaceans (whales and dolphins) in 1953 at Marine Mammal Productions in St. Augustine, Florida. Although the field of animal training has been around for much longer, countless advances have been made over the last 46 years.

Due to the unique characteristics of an aquatic environment, training cetaceans focused largely on a method called "positive reinforcement," the principal means used in conditioning Keiko when he was younger. While Keiko's current goals have changed drastically from what he learned at an early age, we still use positive reinforcement to help influence behavior throughout his rehabilitation.

Goals such as learning to capture and eat live fish, physical conditioning (exercise), swimming through a gate, following a designated boat, and adapting to a new environment all represent significant change for Keiko. Positive reinforcement helps Keiko adjust to these changes by carefully introducing each step and focusing on his successes, one at a time.

Doff and Don

Remarkably, it did not take long to design and fabricate the gateway for Keiko's access to the med pool (despite all the activity surrounding the ongoing barrier net work). Brian and I had drawn out a napkin design of a square aluminum frame, ten feet wide by fifteen feet deep, which would be tied into the medical pool's net wall just below the surface of the water. Once affixed to the net wall, we would cut the center out making a hole in the med pool net framed by the metal outline. In order to open and close our makeshift gate, the two sides of the frame had vertical bars or tracks welded on and running unobstructed up the length of the frame. A free-floating horizontal bar, attached to the frame's vertical bars on each side by eyebolts, allowed the gate to be opened or closed.

The frame was fabricated by a welder in town and brought out to the bay pen by Michael and the valiant Marine Ops crew. Michael and the guys helped us lower the frame into place and anchored it in several spots on the med pool's net wall. The rest would be up to Brian and me. Marine Ops had more than they could handle and could not spend the time tediously stitching the frame into place. As in most animal environments where personnel are divided by specialized responsibilities, the Behavior Team was viewed as soft-handed wimps by the leathered and tough Marine Operations gorillas. Brian and I wouldn't have any of it. We would have this gate installed and in operation in no time and without asking for any help, come hell or high water.

The high water came when we were diving on the outside of the bay pen between the net walls of the north pool and the medical pool. Roughly two feet separated the opposing vertical walls providing only enough space to turn around while clad in full dive gear and a restrictive dry suit. At first both Brian and I had been diving. Monotonously, with hands numb from cold, we went square mesh by square mesh tying the net to the metal frame of the bay pen's middle structure. Eventually, we ran out of dive tanks and had only enough to allow one of us in the water at a time. The other remained topside, warming his core while waiting for the agonizingly slow compressor to refill the next series of tanks.

On one such interval I had been down below and in between the two net walls. Lost in thought born of repetitive work, I was alone in the serene, albeit frigid, underwater world. Void of sound and gravity, a work-dive can be a nice escape. Of course it doesn't always go so smoothly. As was the case on this particular occasion, a swift surge current came out of nowhere and swept the two net panels upward toward the surface, sandwiching me in the middle. Having been a frequent diver, and only at a depth of twenty feet or so, I was not immediately alarmed and resisted the urge to tense my body. The surge came into the bay traveling north, paused and then returned south and out of the bay. As quickly as it came it went. In its wake I was left tangled in a heap between the two net panels.

If pressed, I do not think I could come up with any scenario imaginable where panic would actually be beneficial. It certainly was not in this situation. Although immobilized as if caught in a giant underwater spider's web, I at least kept my cool. For the time being, I had plenty of air—about a quarter tank (in that temperature, approximately fifteen minutes). At first I tried to reposition myself to get a feel for where all the hang-ups were. But it was nothing doing. The snags were behind me. No easy out this time. I would just have to take all the gear off underwater.

In diving, this is called "doff and don," meaning to remove the dive gear and put it back on while keeping the regulator or mouthpiece in your mouth (never losing the air supply). In the process of getting a diver's certification, this is a required skill. Only I had to modify the exercise slightly. Because I was tangled with the first stage caught in the net behind my head and my feet stuck through the opposite net, I had to completely remove the dive gear and take the regulator out of my mouth (though I maintained a death grip on it the entire time).

The maneuver only took maybe a minute; however, the exercise was complicated by the fact that I could not feel my feet or hands. Visibility was only a meter at best. Nonetheless, I managed to get loose from the net, and with equipment in tow, dove down and out of the confines of the pen.

Climbing back on deck, I was pretty proud of myself for not flipping out and turning a minor inconvenience into something more serious. Although emergency skills like doff and don and buddy breathing are practiced, actually *needing* them in a real world setting is quite different than a dress rehearsal. Daylight was growing short and Brian and I decided it would take another day to finish the gateway.

By the end of the second day we had the new gate firmly lashed in place around its entire perimeter—over thirty-five meters of net stitching. After a thorough inspection of the medical pool, we opened the rectangular center of the frame thus granting Keiko access.

The gate acted just like an upside-down window shade. When open, the excess net would sag at the bottom of the gateway. When closed, the bagging net was pulled to the top by the horizontal bar and stretched tight over the opening, thereby closing the gate. Michael, Smari, Greg and Blair all pitched in with final touches and installed a hand-crank system allowing us to open or shut the gate from the side of the pool.

To our amazement, the gate worked beautifully. Gravity did all the work of opening the gate and although shutting it was slow (many turns later on the hand-crank), it operated more smoothly than we anticipated.

OFS Public Web Update: November 14, 1999

> Preparing Keiko for reintroduction involves precision planning. In previous updates we have talked about various goals that must be achieved for rehabilitation. Learning to catch and eat live fish, following a designated boat in the open ocean, and meeting the physical demands of travel in the North Atlantic are perhaps the most obvious; however, there are numerous smaller victories in store for Keiko. Planning for the long-term goals requires breaking those goals down into day-to-day events. It is the implementation of these short-term challenges that becomes so important in setting Keiko up to succeed at the bigger picture.
>
> In last week's update (11/2/99), we talked about the use of DRA conditioning and how it shapes Keiko's behavior. We are currently utilizing DRA conditioning to prepare Keiko for a significant step in the rehabilitation process.
>
> Our marine engineers have been working around the clock on preparations to install the "barrier net," a net that spans 300 meters and will allow Keiko access to the entire bay. Exposure to the bay area represents significant change for Keiko. In Klettsvik bay he will have access to a natural ocean bottom for the first time in 20 years! There is a wide variety of sea life contained in the bay (including some pretty big sea stars). Of course there are some challenges that Keiko must

meet before any of this can happen. In order to access the bay, Keiko must learn to swim through an opening in the bay pen net itself.

To prepare Keiko for this detail of his rehabilitation, our on-site team designed and installed a "gateway" between two pools on the inside of the bay pen. Using this gateway, Keiko will learn to swim in and out of his floating pen and eventually, out of the bay area. This particular opening will act as a prototype for the actual gateway that will be placed in the barrier net between Klettsvik Bay and the North Atlantic.

Gateway Conditioning

Learning to swim through the gate is an important step; however, equally as important to his long-term rehabilitation is "how" Keiko learns this behavior. Each step of the way trainers encourage Keiko to explore his surroundings. Introducing a new item such as the gate offers an opportunity to reward Keiko's initiative.

On Friday November 12, divers completed their final inspection of the new gateway making sure that all was in order and the area was safe for Keiko. That afternoon the gate was opened granting Keiko access to another section of the bay pen.

Once again, it was time for my return stateside for a short spell. In departing, I left explicit instructions for Keiko's conditioning work: It was time to get Keiko through the gate and into the medical pool. It would not be long before we were faced with access to the bay and at present, Keiko was not ready. For starters, he wouldn't go anywhere near the new gateway in his bay pen enclosure. His history with gates was not ideal, and reshaping this behavior was going to require exacting focus. The task fell first to Kelly, Brian and Steve.

Tom

Even when on a home rotation, we were never truly "off." Charles had approved a request to bring aboard one more new staff member. Robin and I both knew exactly whom we wanted. I spent most of my rotation home preparing for the arrival of Tom Sanders. Tom and I had worked together a few years in the SeaWorld of Florida park, which is all it took for us to become close friends. Tom had more aptitude in one finger than most people possessed in their entire being. He was mechanically inclined, adaptive to almost any environment, athletic, and he understood how to shape behavior. He was in many ways a Poet-Philosopher type when it came to the application of behavioral conditioning. These traits along with Tom's affable personality, made him the perfect fit for the project. Where Kelly provided additional behavioral input and expertise, she also contrasted sharply with the rugged conditions. Everyday had to be reinvented, every advancing step a deliberate effort; nothing was handed to you with ease. We needed a pacesetter, someone who created energy where little existed; someone who could lend steadfastness to the otherwise illusory plan of release. Perhaps as important, we wanted someone the existing staff would accept. Tom was our guy. I had great expectations and hopes for his introduction in Iceland. As chance would have it, he would meet me on my return rotation for his indoctrination, and we would spend the first few weeks together with Keiko.

As was often the case in my absence, Robin was consumed with advancing the installation of the barrier net. A critical path to be sure, but it always meant setbacks for Keiko. These setbacks were seldom obvious.

In Keiko's case, we were steadily and carefully introducing immense changes in his life. Some of these changes were an investment in a future phase of the program. The complexity of Keiko's rehabilitation, much like chess, required awareness of steps many "moves" in advance. If this steady progress was disrupted or paused, the eventual effects were akin to taking a cake out of the oven before it has risen completely. The cake will fall before the icing can be spread.

More than a few on the project feigned interest in following protocol and doubted nearly every step absent tangible proof that the tedious demands were yielding results. It is for this reason that behavior is often viewed like religion. This human haste for material and measurable evidence has haunted the beginnings of almost every program in which I have been involved. People are tragically impatient, I among them. But if one persists long enough, ever faithful and steadfast in implementing the principles of learning as we were with Keiko, inevitably something will happen that provides the all-too-necessary testament. It was Tom's initial days on the bay pen with Keiko when this affirmation presented itself.

Returning to Iceland, my route converged with Tom's in the Keflavik airport. Although we had talked abundantly in the past few weeks, we had not seen each other in three years. Without fail, Tom's stubborn resistance to aging did not disappoint. He looked as if it was only yesterday that we had worked together at SeaWorld. Our friendship has always been as comfortable as a trusted pair of work gloves. We fell effortlessly into our usual well-worn jibes and trouble-making rhythms. Throughout the ride from Keflavik to Reykjavik and the short plane hop over to Vestmannaeyjar, we discussed the project, touching on everything from staff to amenities to weather, local customs (the beautiful women of Iceland) and of course Keiko. Tom also knew Kelly well as they had worked together at SeaWorld of Ohio. We shared no small amount of laughs at her expense and favor.

Through e-mails, session records and my daily conversations with Robin, I was already aware of Keiko's failure to go through the new guillotine gate to the medical pool. This despite the singular focus and top priority assigned to the task when I last departed the island weeks earlier. There were many reasons for the lack of progress, not the least of which was the lack of faith in the principles of behavioral modification which in turn placed a deadweight on getting anything done. Kelly was officially in charge of Keiko's daily progress during my rotation home; however, she was often railroaded by the informal chain of command occupying the project at the time. Although she tried valiantly to hold her ground she could not single-handedly be with Keiko 24/7. In her absences from the bay pen, laziness and bold defiance of the conditioning protocols ruled the day. Sometimes conflicting agendas were even carried out in her presence.

Forty-Eight Hours

Klettsvik welcomed Tom with overcast, but mild, weather for his first foray onto the bay pen. For my part, I was as a child filled with Christmas Eve anticipation. Tom's presence was exciting for me, yes, but my excitement was equally enhanced by the fortuitous absence of anyone else on the pen that day. It was no secret that I was disappointed at Keiko's failure to go through the gate. It was the first question I asked each time Kelly and I communicated during my rotation home. Whether the staff was avoiding me or just fatigued from wearisome efforts didn't matter to me, I was happy to have the one-on-one time with Tom just the same.

After a whirlwind tour of the floating facility, Tom and I wasted no time getting to the task. It was our turn "at bat" on the gate training, and both of us were itching for the challenge. That we had just walked into the spotlight with the more resistant members of the staff was not lost on me. Advantage gained by our history of working together, we required little discussion on the planned approach. He and I had been here before. As sudden as if throwing a light switch, we turned off every form of stimulation in the north pool that we had any influence over. We avoided the north pool like the plague, making sure that buckets, tools and equipment of any sort were stationed where we could get to it without being seen by our pupil. We demanded that no boats come anywhere near the bay or bay pen without our express consent. Keiko's only access to us and any form of change in the environment would come from the medical pool via the gateway.

While this approach may seem hard-line—and it was—Keiko's repeated history of failure in gate conditioning produced in him a hardened resolve at avoidance. He had been taught, albeit nonintentionally, to withhold longer and longer in avoidance of the separation. The very act of avoiding something Keiko considered aversive was reinforcing in and of itself. In each past attempt, an eventual return to normal daily interactions in his preferred pool only lent further trophy to his victory. Not this time.

In Keiko's case, we had little choice, and second guessing was not an option. Overcoming his fixation with avoiding gates was paramount for the release process to move forward. Seamless proficiency at gating was required. Time was not on our side. If they were even an option—which they were not—no amount of blue Boomer Balls, toys or antics motivated Keiko to face his nemesis. Likewise, we could not provide his food in the north pool, where he wasn't supposed to be, without also compounding the problem. Tom and I set the stage for ample opportunity, but only when Keiko showed us progress would he also receive the world in all its glorious variety. Starting then, his "world" emanated from within the medical pool itself.

By noon of the first day, we had knocked down nearly twenty individual micro-training sessions. Each and every one of them the same: we stepped up to the medical pool side of the open gateway and tapped the water calling Keiko to position inside the medical pool. As with all conditioning sessions, we waited patiently, first observing Keiko and looking for the ideal time to call his attention. It goes without saying; presenting the signal (stimulus) calling him to position in the medical pool while he was stationary and facing away from the gate would not produce the desired result. But, we didn't just sit around waiting for the ideal happenstance opening, we also created "setting conditions" to encourage him.

Setting conditions are exactly what they sound like. They are conditions that create the likelihood for certain behaviors to occur, such as Keiko moving toward the gate or showing interest in the med pool. It was still early in Keiko's release conditioning. At this stage, his human counterparts remained a bright star in his night sky. The magnetism of his relationship with humans meant Keiko wanted to be near us. Therefore, we moved ourselves and every form of activity to the opposite side of the medical pool.

Tom Sanders atop the cliff overlooking Klettsvik Bay. Keiko's bay pen is shown here after the south pool was shortened due to storm damage incurred in late 1999. A portion of the barrier net buoy line is visible in the middle of the picture. The North Atlantic lies beyond. Photo: Tracy Karmuza McLay.

Keiko's defining three dark spots are clearly seen as he spyhops to watch a cameraman just out of view while the marine operations team in the background repairs a section of the bay pen. Photo: Mark Simmons

View of the town on Heimaey Island from the fourth floor of the staff hotel. The roof of the solarium is visible in the right foreground and a dormant volcano in the background. Photo: Mark Simmons

From left to right: Brian O'Neill, Greg Schorr, Mark Simmons and Smari Harðarson enjoying morning coffee inside the research shack after training sessions with Keiko and a dive inspection of the bay pen. Photo: Robin B. Friday.

Back row from left to right, Mark Simmons, Gudmundur "Gummi" Eyjolfsson, Charles Vinick, Smari Harðarson, Blair Mott, Keiko, Jean-Michel Cousteau, Stephen Claussen, Brian O'Neill, Jeff Foster, Tom Sanders and Robin Friday. Bottom row left to right, Michael Parks, Steve Sinelli, Tracy Karmuza McLay, Kelly Reed Gray, Jen Schorr, Greg Schorr and Hallur Hallsson.

Mark Simmons and Kelly Reed Gray finish an exercise session with Keiko during strong winds and rain in the bay pen's north pool. Photo: David M. Barron © Oxygen Group

Keiko doing a side breach in the north pool. This picture was taken not long before bay access, after many months of physical conditioning work. Keiko had become lean and muscular by then. Photo: David M. Barron © Oxygen Group

Tom Sanders and Mark Simmons on top of the research shack reinforcing Keiko for exploring the bay. Tom is pulling back on the Herring Delivery System (HDS) slingshot preparing to release a barrage of "Herring from Heaven." Photo: Alyssa Simmons

Left to right: Stephen Claussen, Kelly Reed Gray and Mark Simmons take Keiko's routine body measurements. Photo: David M. Barron © Oxygen Group

Keiko's bay pen is shown here during a mild windstorm in Klettsvik Bay. The barrier net buoy line is visible in the foreground. Photo: David M. Barron © Oxygen Group

Mark Simmons and Keiko. Keiko enjoys water spray on his tongue as a form of reinforcement during an exercise session in the bay pen. Photo: David M. Barron © Oxygen Group

Tracy Karmuza stands on the Draupnir's *sponson as Mark Simmons works with Keiko during boat-walk training in Klettsvik Bay. Photo: Robin B. Friday.*

Keiko and the Draupnir *at sea during a boat-walk. Kelly Reed Gray sits on the engine cowling aft, Stephen Claussen and Jen Schorr observe from on top of the pilothouse, and Charles Vinick and Robin Friday are positioned near the bow of the boat while Mark Simmons works with Keiko from the walk platform. Photo: Steve Sinelli*

Keiko looks at author Mark Simmons during an ocean walk in the North Atlantic. The photo was taken from the deck of the Draupnir. Heppin *is visible in the background. Photo: Steve Sinelli.*

Keiko, severely cut and scraped, is shown here approximately a week after his first encounter with ice floes in the fjords of Norway. Photo: Zsolt Halapi

Children in Halsa, Norway get into the water with Keiko. Photo: Per-Tormod Nilsen

Keiko's burial site in Taknes Bay, Norway, photographed not long after his death. Photo: Jan Vimme

Another "lever" that required some pulling was the use of differential reinforcement techniques (Differential Reinforcement of Alternative behavior or DRA). A mouthful to be sure, but also an aspect of Keiko's conditioning that would soon become a primary tool in his preparation for the North Atlantic.

DRA is a simple concept not so simply applied, both timing and frequency determining success or failure. It is a method commonly used to reduce or eliminate unwanted behavior. In this case, we wanted to reduce any activity that resembled gate avoidance, such as hiding on the opposite side of the north pool. In contrast to avoidance, we could reward Keiko for being near the gate, looking above the surface at our position inside the medical pool or a host of other behaviors. Simply put, anything where Keiko faced his fears.

Applying DRA in the narrow context of gate training, we sought also to avoid creating frustration and forestall any potential that Keiko might completely shut down. It was only noon, and we had already changed nearly every rule he was hopelessly accustomed to when it came to gates. Too much too fast and even the most zealous pupils experience a loss of will.

Keiko was not new at this pretend gating exercise. It didn't take long before he was reluctantly poking his giant head two and three feet through the gate in return for the old familiar sound of the whistle bridge, only to then paddle his disproportionate pectoral flippers backing himself out of the gateway. We couldn't get him to come any farther than this bogus three-foot fake without giving him a clear goal that forced him to come farther through the gate each time. Usually that goal was a hand target. However, neither Tom nor I were able to reach the exact spot we wanted. There are limitations to applied conditioning, and they are usually related to logistics.

Initially, we employed the use of the target pole presenting a clear prompt for Keiko to touch with his nose, making inch-by-inch progress through the gate. But when we used the target pole and followed each success by tossing Keiko a few herring, the current swiftly took the fish back to the north pool. Keiko effectively received his reward in exactly the wrong place and for exactly the wrong behavior: for moving backwards. The solution was both fun and frustrating.

Tom donned his splash suit borrowed from Stephen Claussen and stood at the ready, almost too eager to be back in the water with a killer whale after three dry years. "Do you want me in before you call him?" he asked, target pole in one hand and a bucket of fish in the other.

"Let's see how it goes. I think this first time you should just be at the poolside and ready to get into position. I'd rather get a response from him first before we make a change in the environment."

Interested in what we had in store next, Keiko circled the north pool passing by the gate with one eye raised above the surface, peering in our direction. It was a casual glance, one that says, "I see you, but I'm busy at the moment." I waited until he passed. On his next turn toward the gate, I slapped the surface of the water just inside the med pool. Keiko turned in a beeline toward the gate, never altering his pace. Tom slipped in the water and moved to position at the gate about four feet inside the med pool. He placed his hand out about a foot or so below the water's surface, palm outward toward Keiko. Tom kicked his feet fighting to remain upright, his whistle bridge gripped between his teeth. Keiko came straight away and touched Tom's hand without breaking stride. Tom bridged but simultaneously prompted Keiko to remain touching his hand, so he could offer a couple herring before the backpedaling that Tom knew was coming next.

"Tom, here . . .," I said with some urgency as I tossed a herring. As athletic as Tom was, even he couldn't catch the sorry excuse for a toss I had let fly. The herring overshot Tom, passing overhead and beyond his reach. The current quickly took the fish right into the north pool. Keiko swiftly backed out of the gate in pursuit.

"Damn it, damn it, damn it. That was entirely my fault," I said, completely frustrated. I knew I had gotten ahead of myself and thrown the fish, not paying attention to accuracy, and when Tom wasn't ready. "Okay, stay there and call him over yourself. I'll get it right this time. Sorry."

Tom thought better. "Just toss me one now," he said as he slapped the water, holding his other hand expectantly above the surface toward me. "This is an easy one. He'll do it."

Without saying as much, Tom was referring to the very bad idea of having food in your hand while in the water with a killer whale. Had it been any other killer whale in the world, we would have never gone out on that limb. I tossed him the herring, this time without incident.

The bridge came, and Tom handed Keiko the herring in one swift move. But Keiko didn't back out. Instead he sat motionless with his head well inside the gate.

Tom reacted, "Toss me another." It came out muffled through teeth gripping his whistle bridge with no small amount of lisping. I started laughing, both from the success of it and how stupid Tom sounded and looked, jamming his face into the water to see Keiko just beneath him. In the midst of it, he chalked up one more win for the med gate as he gave the herring to Keiko. This time Tom took a couple dog paddles and got out at the side of the med pool.

"Crap, that's cold," he said, briskly sweeping the water from his military-style buzz cut. The water temp was hovering around forty degrees that day, warm for winter.

"Yeah, lot colder than Shamu Stadium. Wanna leave it for a while, let that marinate?" I asked.

"I kinda wanna go back right away. I think he'll come in further this time."

Not one to stand in the way of confidence or overreaching trainers, I agreed. Tom slapped the water. At Keiko's reaction, he immediately got back in. This time he had two herring in his left hand to start.

"Right here? I can't tell where from down here." Tom asked, trying to adjust his position further into the medical pool.

"Back a little, that's still pretty close to the last. I'd ask for at least a foot more and see what he does. But if he gets pissed and comes into the med pool to chew your ass, I'm bridging it . . . just so you know."

Tom ignored my jest or at least couldn't respond right away. He was too busy fighting the northbound current that had recently picked up. The current wanted to sweep him out of the med pool and into the north pool. He looked absurd. Every time he put a hand forward as a target for Keiko, he lacked the ability to hold his position, drifting into the whale. A one-armed man sculling this way and that. Together we must have looked like two of the Three Stooges—the third a few feet away in the gateway.

After a few more attempts, we called a time-out and regrouped to find a better approach. The conditioning was going well. Keiko was advancing quickly, although we were completely vexed by the elements and the simple task of staying in one place in the water. Not to be undone, Tom and I decided that scuba would be the solution. First, the current was being driven mostly by the wind and therefore should be much easier to deal with a few feet below the surface. Second, placing ourselves even with Keiko in the water column offered him a much more natural entry position (recall that the gate opening was a couple feet below the water). It was worth a try.

This time I would be the target in the water. After gearing up, we waited for telltale interest from Keiko. My first attempt proved out our theory, as it was much easier to be precise in where we wanted Keiko to come. It was also easier to provide immediate reinforcement. The remainder of the afternoon each successive attempt went on the same way. Inch-by-inch we slowly-but-surely gained ground on each "ask." Often in these situations, whether it's the first time an animal has gone through a small opening or like Keiko with a bad history of gating, there comes a time in the approximations when you're in their way. I thought we were there.

"He's past his dorsal fin through the gate, kinda at that breaking point where it's easier for him to come all the way in and turn around rather than backpedal out." We were standing on the lee of the research shack in the sliver of sun left, attempting to warm our bodies. Through chattering teeth I said to Tom, "You need to be close to the poolside. If he comes right to me this time I'll bridge and move aside. Then you slap. I think we're holding him back. At this point he can just as easily reach you as come to me. I think he's almost there."

As proposed, we set up the scenario. Keiko came directly to my underwater hand-target with little hesitation. I bridged, clumsily taking my regulator out and blowing the whistle underwater, then sculling to the right and out of Keiko's path to Tom. On the surface, Tom slapped, keeping his hand extended below the surface to meet Keiko halfway. Keiko turned his enormous head and torso in Tom's direction, but didn't move any farther. Then, after some hesitation, he committed to Tom, touching his outstretched hand slowly and delicately with his nose. By then I was at the surface carefully backing away. I didn't want to crowd Keiko's entry path or cause uneasiness. Tom bridged. Then Keiko surprised us, he sat up with his head above the surface and mouth open, ready to receive herring, but his body was still in the gate. His entire nether regions and flukes still hung onto the north pool, while the rest of him was in the med pool contorting in order to reach Tom's hand. Keiko was fanning his huge pecs beneath the surface to hold this oddball position. As Tom broke away from the session, Keiko once again backpedaled out of the gate.

"Oh my God, what a freak'n baby!" I said as we both laughed. The progress was good. We could eliminate the tedious scuba step, but expecting that Keiko would "break the barrier" and come fully into the med pool was overly optimistic. Instead, he showed us that he was comfortable with eighty percent of his body inside the med pool, but not so much that he would actually come fully through the gate.

"I can't believe he'd rather back out than just come through and turn around. It takes him almost a minute to get his body out of the gate backwards. He looks ridiculous," Tom said.

"No kidding. And here's the whale we're going to release, can't cope with a simple gate. Watching him paddle his pecs back out of the gate while trying to avoid touching the gate frame, as if it's going to bite him, is pretty damn funny. I've never seen a killer whale act so chicken-shit." One of the benefits of my solitude with a trusted friend is that I could speak openly, even if a bit dramatically for effect.

Without a word, Tom stepped back up to the same spot at the med pool, just to the side and about six feet from the gate. He was taking advantage of momentum. Keiko responded to his slap and came partway through the gate, gingerly twisting up through the opening to ever so slowly touch Tom's outstretched hand. The bridge was given, and again Keiko sat up above the surface with his mouth gaping wide. He literally looked as if a child expecting great praise for his little accomplishment *Ta-da!* I had to remind myself this was a killer whale.

The day was closing fast. We completed a few more repetitions, each time requiring just a frog's-hair more of Keiko. Tom worked his way down to the right of the adjacent med pool wall, never once repeating the previous step and stretching Keiko's limit on remaining in the gate. By our last session of the first day, Keiko was in the med pool with only his flukes hanging through the gate, yet each time he still chose to back himself all the way out. At this point, backing out was a process that took him considerable effort, contorting and twisting this way and that. Amazing how this enormous lug of an animal could navigate backwards through the gate without even coming close to touching the gate itself.

On the second day, Tracy accompanied Tom and me to the pen. Knowing them both, I knew this was going to be a fun trio. Tom and Tracy were fast friends. There was the usual formal exchange here and there, probing questions shrouded within small talk, but very soon thereafter they were chiding one another as if old compatriots. Of course I didn't hesitate to encourage such behavior. After filling Tracy in on prior successes, we jumped right to where we had ended the previous day.

Regression is a normal part of learning. Often such a bold attempt to pick up where we left off would only be met with failure.

Failure, however, is also an important component of learning, providing a juxtaposition to success. But we were fortunate this day because Keiko was up for the challenge. He came right to Tom's position in the med pool on the first try. Likewise, he also continued to hook his flukes back through the gate reversing himself out at the end of each attempt.

Following a brief discussion, we agreed it was time to deliberately create failure. In the next approximation, we were going to ask Keiko to come so far that he would be unable to keep his flukes in the gate. A mere two or so feet further down the med wall would do the trick. This time, Tom surprised Keiko by picking him up in the north pool, the first time we had acknowledged Keiko's presence in the north pool in the last thirty-six hours. Tom got exactly the reaction he was looking for: Keiko popped up with greater-than-usual energy and attentiveness. After all, coming over to Tom in the preferred pool was fall-off-a-log easy. Taking advantage of Keiko's five tons of momentum, Tom didn't wait for the Big Man to come to a complete stop. As soon as he had eye contact, he immediately pointed to the med pool, turning and scaling the eighteen-inch step. He ran posthaste to the receiving position in the med pool. Without giving so much as a glance back at Keiko, Tom "acted" as if there was no question Keiko would follow.

It was not to be this time. Keiko came through the gate but stopped short of meeting Tom's position further down the wall. As we expected, this would require him to fully commit to the separation. Tom gave him longer than necessary but finally stepped away from the poolside unsuccessful. The three of us moved away from the pools while Keiko danced his way backwards out of the gate.

Part of the day's plan, we reduced the number of opportunities we gave Keiko to succeed. For the most part, his successes by the afternoon of the previous day were plentiful; he didn't need the high-frequency micro-sessions as in the beginning. Also, it allowed us to put much more of Keiko's primary reinforcement into each progressive step forward. It was time to let him sit for a while and let the failure of the previous attempt take effect. In the simplest form, the more he succeeded, the more he got and vice versa.

This exercise was essentially a balancing act. Waiting too long was also detrimental in that we could lose the momentum of both success and failure. After twenty minutes, we stepped back up to the pools. It was only one p.m., and we had a long way to go. Tom stepped first to the north pool to call Keiko, again pointing him to the med pool at the instance of eye contact. Tracy and I waited near the research shack and med pool, arms crossed and silently watching the new guy work his trade. I enjoyed watching Tom. His bent and expectant stance was a trademark quality from his prior work with the ocean's top predator.

Keiko disappeared beneath the surface, as he had done on each occasion, to enter the submerged gateway. Tom crouched as if a tiger ready to spring, poised to react to Keiko's position and encourage the extra distance. Surprising all of us, Keiko popped up fully in front of Tom, completely inside the medical pool, as if he'd done this his entire life! Tracy nearly burst my eardrums screeching in excitement. Tom was pounding herring into Keiko's open mouth.

"Tom, point him out, point him out!" I yelled.

Tom did precisely that, and Keiko turned slowly but in the wrong direction. He had turned to his right and thus into the smallest corner of the med pool. There was not sufficient room for him to turn his dump-truck sized body. He eventually made his way out but with some whitewater included. It wasn't a pretty exit and risked creating aversion on his first breakthrough into the medical pool.

The three of us made quick adjustment. On the next go-round, Tom received Keiko at the same position, but this time when he pointed him out of the med, he led Keiko to his right—Keiko's left—and the larger expanse of the pool. Tracy stood ready on the opposite side slapping the surface to provide a backup to Tom's guidance. This time Keiko turned smoothly, swimming back through the gate to the north pool. In both instances of these first full separations, it was important that Tom point Keiko back out of the medical pool quickly, before he split of his own accord, gaining his trust that it was not a trap. As he entered the gateway from the med-pool side, Tom bridged again and tossed a single herring in his path, reinforcement for a well-executed departure. We broke from that session and rejoiced. It had been less than forty-eight hours, and Keiko had finally faced his demons.

For the remainder of that day and the following week, we continued to incorporate the gateway into every session, providing the bulk of Keiko's food and fun in the medical pool. Within a few short days we could ask Keiko to the med pool and shut the gate for extended periods with consistent results. We finally had the practical means to gain access to the bay and the North Atlantic beyond. Keiko was ready for Klettsvik.

Although no one person ever gave voice or affirmation, the accomplishment went a long way in fostering relationships with the original staff. In my simple estimation, it was proof positive that we knew what we were doing and could produce results; the testament so desperately needed and timely. Those forty-eight hours achieved so much more than simply reshaping a single whale's dismal history with gates. On the extended eve of the barrier net's installation and a plan forward, it marked the emergence of the true Behavior Team.

Phase II, Klettsvik Bay

The entirety of the Marine Operations team was on-site for the first time since the bay pen was constructed nearing two years prior. Some of them had become so accustomed to working apart from each other on opposite rotations that getting together under one roof, so to speak, became as uncomfortable as a family reunion. Both excitement and uneasiness hung over the water.

Among the mild contentions, Greg Schorr and Michael Parks each vied for lead dog of Marine Operations. Normally, they served on opposite rotations. Greg was so affable by nature one would have to make a considerable effort to alienate him. Michael, on the other hand, had a way of getting under the skin of his workmates. Nonetheless, Michael was the senior of the two and boasted easily twenty years on Greg. Despite his offenses, unintended as they were, Michael's experience was highly respected. He was "captain of the ship."

Somewhere in his early forties, Michael's full head of boyish black hair, easy smile and frequent "knowing" smirk, along with a drawl when he spoke, set forth the traits of a western rancher. But it didn't take long working with Michael to recognize the salty dog sea captain underneath. He had wisdom in his eyes, the product of seasoned experience gained on the decks of Alaskan trawlers. On-site in Iceland he lived by himself, apart from the other expatriates on the team. In his own distinctive way Michael was a loner.

During the installation of the barrier net, activity surrounding Keiko's bay pen abounded. Our challenge was to take advantage of the downtime in between work hours, the limited windows where we could avoid direct association with so much human activity. As weather was the antagonist to Mighty Mo, so too was the constant presence of Mighty Mo to Keiko's conditioning goals.

Hell Hounds

The following provides perhaps the most succinct description of the barrier net particulars: High performance polyethylene netting technology, 283 meters long, sixteen meters deep. More than forty rock helical or spiral anchors embedded, grouted into rock and epoxied. "Big ass" chain running the length of the bottom-line from east to west, eight to eleven deadweight anchors on the northern and southern sides and nearly 200 individual 350 kgs lift-capacity buoys affixed along the surface line.

The concept of putting a "gag" on the mouth of Klettsvik Bay and her gusty, wild temperament was laughable to most every engineer interviewed during the pursuit of the outlandish solution. Nonetheless, an animal need dispossessed by technical challenge was a combination Robin could not resist. His stubborn persistence that it could be done transported the barrier net from concept to reality. Through Woods Hole ingenuity and a host of hardened Nordic contractors, installation of the net finally began in earnest late December 1999.

Over the course of eight weeks, Mighty Mo labored. Under pressure of limited weather windows, availability of working vessels—in some cases makeshift equipment—and back-to-back–to-back dives, the team struggled forward.

Certainly the mathematics of the barrier net's design was of paramount importance, but no amount of immaculate engineering could replace the dedication of Mighty Mo to effect its installation in the predictably unpredictable conditions. This crew of Marine Operations personnel had been through hell and back, compliments of Klettsvik's vile temper. More than anyone, they truly understood the extreme tests the barrier net would have to endure. They also took to the task of its installation without pause. There was no shortage of the impossible, improbable and downright ludicrous tasks thrown at the team day, night (mostly night) and amidst the worst of conditions. At times it seemed Hell Hounds mercilessly terrorized the effort as if we were attempting to close the gates of hell itself.

Proud or stubborn, perhaps both, Mighty Mo wasted no time on conjecture about whether the job could be done, only *how* it would be done. They had become so adept at the art of problem solving; not one of them realized how well they worked through myriad issues facing the installation. Only a bystander could appreciate the tenacity with which they erected this giant underwater sail.

Much of the work was made possible from the favorable deck of a commercial catamaran called the *Hamar*. Her aft deck almost the size of a tennis court, she was well equipped for the task. Heavy lift davits, wenches and air compressors lined her gunnels in the working area. The cabin and pilothouse were reminiscent of a well-used machine shop, a grimy luxury after the team completed long nitrox dives and took the occasional break to thaw their bones. (Nitrox is a gas mixture of nitrogen and oxygen used to extend dive time, usually in commercial diving operations.)

Winter water temperatures reached as low as thirty-six degrees Fahrenheit. Even their dry suits thickly layered underneath could not resist the sieging cold for long. Each of the Mighty Mo team members took his turn, each pushing himself too far. No one wanted to be "that guy," even though the weight heaped on his shoulders was far from reasonable. The deadness of cold seeped into their bodies so slowly, so surely, that when their hands stopped bending to their will, the exposure had already taken its toll. The core of the body becomes so desperately frigid that it literally takes days to recover.

Such were the ongoing conditions throughout the installation of the barrier net. Affixing the net to the rock walls and sea floor was finally completed in early February 2000, but the barrier was not yet functional. Much like the construction of a building or house, the big work completed left behind a tedious list of finishing touches.

The surface line of the net spanning almost 800 feet across the bay had to have a visual barrier above the surface. This meager addition announced the presence of the net to unsuspecting boat traffic. As importantly, it would deter Keiko from pushing the top line down and swimming over the net, a skill he had learned long ago. Another significant modification, authorized boats needed to be able to traverse the net for passage to the bay pen and the almost daily task of bay pen maintenance. The solution involved a boat gate consisting of several solid foam core cylinders conjoined at each end with eyebolts. Their centers spun on an axis, thus permitting boats to power up and over the buoyant black Caterpillar gate, as they gently rolled beneath conveying a rather peculiar waterborne "speed bump."

Over the course of the barrier net's existence, this boat gate would hold up marvelously, working almost to perfection. But not every boat in our collection was fit for the crossing. Mastering the entrance to the bay in the smaller and lighter *Sili*, equipped with a single outboard motor, was a spectacle to observe. Getting it right required an exact amount of ramming speed timed with lifting the prop clear of the water just before hitting the boat gate. Too little momentum and the boat ran aground on the black cylinders, bow down and prop impotently clear of the water. In weather, this routine was at times welcome entertainment for spectators on the bay pen awaiting their replacements (and no less so for Keiko).

By mid-February, the barrier net had become a reality. What would come next seeped into the consciousness of every soul onsite with finality. Expanding operations to Klettsvik itself constituted the most significant step toward release since Keiko's move to Iceland more than a year and a half earlier. No one had stopped long enough to absorb what the barrier net represented, so lost were we all in the hectic battle to accomplish the impossible. Returning staff were immediately taken by the sight of it, but more so by the way it defined the bay. The enormity of the new enclosure seemed staggering, its breadth and width dwarfing the bay pen and balancing what was previously so inadequate amidst the overreaching heights of the sheer rock walls surrounding Klettsvik.

The achievement breathed new life into Keiko's rehabilitation and the release team. People were starting to believe in each other. The culmination of Keiko's day-to-day progression combined with conquering the closure of the bay was an awakening. In the simplest form, there was a solid release plan, and we had proven unwavering in our pursuit of that plan. Our emergence into Klettsvik Bay marked the beginning of Phase II and a release plan taking explicit form.

Haddock, Lightly Stunned, Hold the Sauce

In the background of activity surrounding the barrier net installation, the Behavior Team began the first steps of live-fish conditioning. At first, the goal was merely to confirm Keiko's willingness to eat live fish. Thereafter, we planned to incorporate live fish into his diet on a random schedule. Contrary to popular belief at the time, we had no intention of transitioning to a complete diet of live fish, or even providing it with any regularity. The topic had long been debated within the release effort. Some equated Keiko's eating of live fish with survival, a singular assurance that he could exist on his own.

Robin and I hummed to a different tune. Killer whales, like many other socially complex mammals, forage and eat cooperatively. The aspect of eating live fish was to us nothing more than a novelty that had to be introduced and rehearsed until it became a familiar part of Keiko's life. We argued, somewhat emphatically, that successful integration with other whales would introduce the opportunity to forage; that observational learning would prevail; and that Keiko need only have experienced the odd feel and texture of many types of live fish beforehand. Regardless, the idea of live fish as the be-all, end-all criteria for release never fully disappeared from the conversation. This particular component of survival had been so overplayed in the media that the organizations responsible could not avoid the ranking question.

Never one for being shy in sharing my opinions, I challenged the FWKF's decree that eating live fish was a decisive factor in Keiko's survival. I called into question the folklore that Keiko had eaten live fish in Oregon and again early after his arrival in Iceland. Based on everything I had seen and knew of Keiko, his disposition, the trials we ran, and stories abundantly offered by the original staff, I knew the prospect that Keiko had hunted down and devoured live fish was highly exaggerated and done so to satisfy donors (or as likely to dispel detractors). It's one thing to eat a disoriented and lethargic fish and entirely another to chase down a dinner that has its own ideas on survival. The distinction of how *live* the live fish actually were was never expounded on in the public arena.

On one occasion, I was assigned by Charles to do an interview for CNN Europe. The suggestion was made that I answer the inevitable question affirmatively, that Keiko was indeed eating live fish. During the actual interview, I sidestepped the issue providing more explanation of the unimportance than any direct confirmation on the subject. It was the last interview in which I was asked to participate.

After extensive over-planning, we finally had a means to keep fish alive on the bay pen and a source to provide the sacrificial haddock, cod or pollock with regularity. Initially we introduced the "not-so-live" fresh fish at the expected time of reinforcement when Keiko normally received thawed herring. After he ate the new fish he immediately received his familiar diet, thus reinforcing the new food type. Gradually, we left the live fish more and more lively, exercising our skill in variable levels of fish stunning with a rubber mallet. It was an imprecise method at best. More than a few foggy fish made a fast getaway when plopped unceremoniously into the bay next to the giant predator. Keiko eventually convinced me that he was willing and capable of eating the wiggly fish, but he never exhibited a predatory response, even when given ample opportunity. He preferred his haddock lightly stunned.

Keiko was given the dazed fish for a short period of time, mostly during his tenure in the bay pen and early days in the bay, but always in the company and supervision of his trainers. The context in which he would *take* live fish became an important distinction. The animated supplement even reached nearly forty percent of his intake for a spell, but this was the only time in the release effort that live fish played a daily role. Much more daunting release objectives demanded ours and Keiko's undivided attention.

In preparation for the next stage of release, we began to shift more and more of our attention on Keiko's time away from humans and less on direct training sessions. Our direct interactions were now reduced almost exclusively to required husbandry. His only other vice, exercise sessions, were orchestrated at lesser intervals; at least until we had other means to keep Keiko active outside of the bay pen. We often purposefully created activity on the decks of the bay pen, deliberately making sure to offer no response to Keiko, as if he had become invisible to the two-legged land dwellers. Covertly, we lurked in unseen corners of the pen, waiting and watching for active swimming or Keiko's occasional chasing of a happenstance seabird that had alighted on the calmer waters of the pen's interior. Nonetheless, novel events were hard to come by. Keiko's activity increased, but this mainly consisted of constant swimming. There was little else to fill the void.

Enter the *Draupnir*

Once in the bay, a fundamentally important step would be introduced: training Keiko to follow a special purpose walk-boat. After toiling over the ideal candidate to serve as Keiko's guide, Michael located a retired coast guard boat aptly named the *Draupnir*. Project leaders approved the acquisition with little hesitation. The *Draupnir*, from Norse mythology's "the ring of Odin," had a complex role among the gods. Long removed from its heyday, the *Draupnir* required no small amount of TLC in preparation for her guardianship over Keiko. The boat had been swamped at least twice in its storied past and, as a result her aft inboard engines were rife with mechanical problems.

The pilothouse, a white metal box located amidships and framed by the bright orange of the foam-filled sponsons, contained just enough room for three to comfortably stand, a fourth if he or she wasn't shy. To the right of the helm wheel were two throttles like that of a commercial aircraft, one each for the starboard and port engines. *Draupnir's* special jet-driven power came from engines that sucked water through an intake. Internal propellers then vigorously jetted the water back out through directional scuppers, thus providing not only movement, but movement in any direction. A dial allowed the captain to direct the boat's propulsion. The *Draupnir* lacked external props that would otherwise serve as underwater blades posing a threat to Keiko.

In addition to twenty-six knots all-ahead, she could "walk" backwards and sideways or hover broadside in a strong current. The *Draupnir* was an ideal escort for the journey ahead, if only she would run. Neither the crew nor Keiko would be well served by an unreliable boat in the midst of the North Atlantic's teeth. Justifiably so, much resource was put to the task of mechanical refurbishment and little toward the unsightly battle scars that revealed the *Draupnir's* hard life. After all, if the release effort went according to plan, and once her role as the walk-boat played out, the old girl would be sunk offshore never to influence Keiko again.

While the *Draupnir* remained nestled in the harbor under the urgency of repair, preparation for Keiko's exposure to the bay drew near. E-mails shot between Santa Barbara, Seattle, and Iceland with increasing necessity and intensity because of the major PR event that was about to unfold. Keiko's initial exposure to the bay was enticing not only for those of us on-site, it was as much a desperately needed storyline for administrative personnel constantly struggling to shore up donor support. Keiko's first foray into the bay was exciting to be sure, but also carried with it an expectation of shock and awe. Many anticipated that a vibrant whale would triumphantly charge from the confines of his pen, as an inmate might run from solitary detention to the fresh air and sunlight of freedom.

Everything we knew about learning suggested otherwise. As with every other systematic change we introduced, the shift from the confines of the pen to the relative vast expanse of the bay would be nothing more than one more approximation in Keiko's rehabilitation. How that change was introduced was critical. There would be only a limited set of opportunities for us to practice transition to the open ocean, albeit, on a much smaller scale and in an environment we could control (to certain extent). The FWKF board, Charles and even some within the staff assumed that the larger bay would naturally be positive for Keiko, that the size alone and the varied stimulation it provided would win the day. However, there was no basis for expecting the bay to be either directly or intrinsically appealing to Keiko. He had no history with such an environment which, after all, was counterbalanced by a long-standing and vast history in smaller and more familiar surroundings.

March 2, 2000, the day before opening the gateway to the bay, Tom was on-site in charge of Keiko's daily management. In anticipation of the transition to the bay, I had seized the opportunity to spend a brief two weeks at home in Orlando. Back in Iceland, every available body worked to clean up the bay, a catchall that often consumed large amounts of trash coming from the shipping channel. As the last of the details were completed, the draw of the public relations event became too enticing to await my return. Keiko's trial release from the pen would move ahead without me. Tom and I exchanged constant e-mails and racked up a hefty phone bill discussing the ideal course of action to be followed.

"How's it going up there?" I started. My morning his afternoon, we were separated by a four-hour time difference.

"Weather's been unusually calm, so that's good. Greg and Blair just finished taking the last trash bags off the beach this morning, so I think the bay cleanup is about as good as it's going to get."

"What about the barrier net, did they get the boat gate working?" I knew before answering that Michael was never satisfied.

"Ha . . . you know Michael. It works really well in my opinion, but he's been messing with it a lot and taking boats over it constantly . . . made it difficult to coordinate sessions with Keiko. Been a pain in the ass, really. Keiko ignores most of it, but he's been logging a lot more lately so that's made it hard to get any DRA in . . . between that and the boats."

I wasn't surprised. "I'll bet you anything that he is watching more than we can see. If he's logging more at night and in the down times, when there's no activity in the bay, it tells me he's tired. Ya know, just from the hubbub, the constancy of it."

We had seen delayed effects of environmental stressors before, sometimes two or more weeks after the fact. I knew that Robin was monitoring this aspect carefully, mandating scheduled down time at changing intervals. Still, the race to the finish was too alluring. Work intervals had increased in duration and frequency over the last week of preparation.

"He's eating okay, and sessions are about the same as when you left. He's just logging more in between," Tom replied to my guess.

"Have you talked to Robin?" I asked, wondering if they had discussed the potential negative impact of over-stimulation.

"I've hardly even seen him. We've been on the bay pen mostly in the afternoon and evening and he's been on the *Hamar* all day." Tom sounded indifferent. "I talked to him over the radio a couple times. He's coming on the pen today to go over the plan."

I immediately became defensive. I knew all too well that Robin already had his own distinct ideas in mind for the bay access. "Listen, this has to be taken very slowly, I know you know that, but I want to make sure we're all on the same page. We can't afford to miss this opportunity. I've copied Robin on the e-mail I sent you yesterday, so I'm hoping he got it . . . and not when his eyes are glazing over at the end of the day."

"I don't know, but he'll be out here today. Maybe we can call you from the bay pen?" Tom offered.

"The most important thing is that we don't rush the process. This is the ideal dress rehearsal for open ocean work. I want Keiko to take the initiative. Just open the gate and break, then standby and be ready to reinforce him when he shows interest in the gate or if by chance he goes through on his own," I replied.

We'd been through this before. I repeated myself as much for my own comfort as to communicate what Tom already understood. "I was going to call him to the north pool, separate him to the med pool, open the bay gate and break from there."

Without needing to say it, Tom knew not to shut the medical pool gate thus cutting off Keiko's retreat to the familiar north pool. His intention was to simply place Keiko close enough to the bay gate removing any doubt that Keiko was aware of the new opening.

"Perfect. I'd station Tracy and Brian outside the pen or on opposite sides of the south pool to reinforce if he comes out on his own."

"Yeah, we talked about that yesterday after you sent your e-mail." By now Tom was just placating my need to be heard. I wasn't telling him anything he didn't already know and know well.

Driving the point home, I closed the topic, "I don't care if it takes all day or two or three days. Just stick with that plan . . . and call me if you need back up."

Perhaps overbearing in my approach at times, I was hell-bent on executing each and every minuscule step to perfection. This was a big one, which only added further intensity to my otherwise incessant brow beating on the finer points of behavioral sciences. One all-important aspect of the new Keiko we needed to find was his inner extrovert. We needed to encourage, in fact shape, an outgoing and curious bull killer whale. A collateral benefit, we had a very rare opportunity to provide positive consequences to enormous environmental change. Environmental change was the central theme of Phase II and III in the release plan. As it was, human haste and the constant need for immediate satisfaction ran in completely the opposite direction.

Enter Keiko

Fortune smiled on the media interests. The weather on March 3, 2000, was impeccable, setting the stage for a seemingly momentous occasion. It was midday, the only time in the short days of winter that provided enough light for the event. Robin, Tom and Tracy were on the bay pen to work Keiko. As they later related, the *Draupnir* had a front row seat, just outside the barrier net, cameramen on her bow. Other members of the staff and media perched atop the overlook on the southern shores of the bay, snow covered volcanic rock framing the scene. All waited for the singular decisive unveiling of Keiko's glorious emergence from the pen.

On the pen, Tom stepped up, waited for the right moment, then called Keiko to the north pool platform where he could get down at the whale's level, even with the water's surface. Keiko popped up, his giant black and white head divided by the pink of his gaping mouth. Tom blew his whistle bridge and moved swiftly to the west side of the medical pool nearest the bay gateway. There had been little activity leading up to this point, and Keiko was excited to see Tom. The whale broke sideways so quickly that he cast a sizable wake as he moved to follow. He hadn't sat up in front of Tom more than a moment when Robin and Tracy opened the doorway to the brave new world. Keiko didn't even flinch. It was as if the most sensory equipped predator was completely oblivious to the clanking clumsy opening of the makeshift gate. Wanting to save most of Keiko's food that day for his hopeful entrance into the bay, Tom tossed him a single herring and stepped back, starting toward the new opening in the pen. Tom joined Robin and Tracy near the entrance to the bay, a steel bucket of fish by his side. The three waited and watched with anxious anticipation.

At first, nothing. Keiko had disappeared from sight. They scanned the interior of the bay pen and finding nothing, stole an occasional expectant glance outward to the bay. Still nothing. After what seemed an eternity, Keiko finally surfaced. Although he had divulged no initial awareness of the new opening in his floating pen, he was indeed keenly aware and wasn't at all comfortable with the strange arrangement. When he surfaced, he had come to a complete stop, logging at the farthest corner of the north pool and facing completely opposite the bay opening. He appeared as if a child pouting in the corner, unwilling to face his parent.

As they waited and watched Robin responded to nagging inquiries emanating from the handheld radio, on the other end of which was Hallur Hallsson, the Keiko Release Project's director of communications. Hallur was attempting to stave off media impatience with an educated response. Initially, Robin had given the order to leave Keiko alone, wait him out. Keiko on the other hand didn't wear a watch and had no obligation to impatient reporters with barren cameras. As minutes bordered on an hour, Keiko sat motionless at his original position as far away from the gateway as he could physically be.

Robin finally had enough of the waiting. The light was fading fast, and he was under pressure of expectant eyes. "Tom, call him back over. . . ."

"You don't think we should wait?" asked Tom. It was more of a statement than a question.

Ignoring it, Robin continued, "Maybe ask for a few behaviors and then let's point him out."

His head cocked slightly and chewing on his whistle bridge, Tom asked, "You want me to wait until he's swimming or call him from there?" His tone ever so slightly contrite, Tom knew Robin's instruction was breaking protocol. We would never provide a rewarding change while Keiko was sitting motionless. Avoidance would get him nowhere in the North Atlantic.

"Just call him over, start in the north pool and move him to the med pool before you point him out. Let's see what he looks like first." Robin gave a pained look. He wasn't oblivious to Tom's undertones.

Knowing that debate would go nowhere, Tom did precisely what Robin asked. Once Tom had Keiko in front of him on the south wall of the med pool, he glanced over his right shoulder at Robin.

Robin was muscling the situation, and he knew it. With purpose, he projected a tone that did not invite question, "Go ahead."

Tom pointed Keiko toward the bay gate and moved immediately to the new platform tied on the outside of the south pool, just to the right and outside of the bay pen. Squatting in his blue and yellow splash suit pulled down around his waist, his knees in the water, Tom slapped the flattest spot he could find between the small caps rippling the surface. Keiko could easily hear the thunderous clap, but he didn't need it. He had already made his decision. As Tom moved right and to the outside of the bay pen, Keiko sank out of sight and went left. He resurfaced back at his position of refusal in the north pool.

After a few moments, Robin told Tom to slap again. Although Tom knew it was futile, it was also necessary to drive the point home. Keiko wasn't going to move. His position taken up at the back of the north pool defiantly communicated a clear message. At Tom's slap, Keiko took a full breath and bobbed his head once, as if he was thinking about moving. The gesture was more likely the first minuscule sign of frustration boiling to the surface.

"I think we're going to make matters worse if we keep asking him," Tom offered.

Robin said nothing. His face stern, he chewed his bottom lip as if he was on the verge of a solution.

Tracy, not knowing this look from Robin, offered observation. "I've seen this before. I don't think we're going to get much out of him for a while." She drew out the first few words, sounding almost jovial.

It was the last thing Robin wanted to hear at the moment. He didn't so much as acknowledge the comment. Tom and Tracy exchanged smirks but dared not say anything. Without a word Robin turned and walked into the research shack.

"What did I say?" Tracy asked.

"I don't know why we're in such a rush. I mean, I know why, but they can wait. You didn't say anything that's not true . . . just not great timing," Tom offered, a useless consolation. "I can tell he's frustrated."

"Robin or Keiko?" Tracy couldn't resist.

"Yeah . . ."

Some time passed, not an hour, but not less than half. Robin had been on the phone in the research shack apparently discussing the reason they waited, presumably with Hallur or Charles. When he finally came back out on the deck, he seemed to have settled on a course of action.

"Okay, Tom, let's go ahead and call him over again. This time call him straight to the med pool. Don't let him stop. Point him out right away."

A five-ton killer whale doesn't start or stop effortlessly. Rather than ask for the energy required to stop and then start swimming again in order to follow Tom, Robin wanted to use Keiko's momentum to advantage and in the direction of the bay.

Again Tom stepped up. Again he followed Robin's lead to perfection. Again Keiko went to his spot in the north pool. But this time it was different. As Keiko was pointed "on the fly" he indeed did move toward the bay gate, at the last moment diving down and twisting in an underwater arch away from the offensive opening. As he surfaced and slowed to assume his position in the north pool, he slapped his flukes—a clear indication of mounting frustration.

"Whoa, he's pissed," Tom pointed out the obvious.

"Good, let him get pissed," Robin said. "Sure better than just sitting there."

"What now?" Tom had completely given over to Robin's direction, not wanting to fight a losing battle.

"Give'em a few minutes or so and try again."

After those few minutes Robin offered another change of direction. "Tracy . . . " A pregnant pause. "You point him out this time, same as before. This time Tom, you get on the platform in the bay and slap as soon as she points."

Turning slightly away from Tom and Tracy, Robin lifted the handheld to his mouth. "Hallur, we're going again." The notice had been given before each effort throughout the morning.

Responding to Tracy's slap, Keiko remained mostly horizontal in the water column as he approached. He knew what was coming, and this was an easier position than turning vertical and upward toward Tracy. As his right eye lifted slightly and made contact with Tracy's searching expression, she pointed him toward Tom and the gateway, then moved fluidly toward the gate as if saying, "Follow me" with her whole body.

Keiko dove below and out of sight. There was no white water evidencing a hard turn away, but he had gone deep enough that not even the glowing white portions of his markings were visible in the tealike depth. On the bay platform, Tom slapped.

Keiko was nowhere to be seen. A minute passed. Still there was no sign of the whale. Robin searched the bay expectantly, encouraged by the prolonged disappearing act. If Keiko was deep and looking at the gate, he didn't want to miss the chance to encourage him.

"Slap!" he yelled in Tom's direction.

Tom had been looking back over his left shoulder as he sat on his knees on the nylon-webbed platform, his right hand steadying the fish bucket at his side. As he turned about face and leaned down on his left arm preparing to slap, Keiko surfaced almost hitting Tom's face with his bulbous black rostrum. Instead of slapping, the athletic Tom smoothly transitioned from raised hand to placing a target on Keiko's nose at once blowing his whistle. A continuation of the same fluid movement, he turned and began grabbing a ball of herring with both hands, triumphantly plopping them with a *thwap* into Keiko's cavernous mouth. Only then did he take the time to give cheer. After all, Tom knew he was on camera.

"Think he'll do a behavior?" Robin was directly behind Tom leaning over the yellow plastic railing. He finally looked relaxed, his faded blue jeans, blue sweatshirt and black fleece vest now appropriate for a more casual mood.

"Only one way to find out," said Tom.

Tom finished feeding Keiko a few generous handfuls of fish, then standing, he bridged again and fed him a second time, as a quick "thanks" for not leaving when he stood up. Keiko's eyes were wide and alert. Clearly his presence in the bay was tentative. Tom shuffled a few inches to his right and made a dramatic sweeping gesture, smacking his left outstretched arm down to his extended right arm, as if a Florida Gator fan taunting their opponent. The signal was asking for a tail-lob.

Keiko moved to his right, lined up at the surface, took one long delaying breath and then began slapping his tail flukes on the surface as he swam a counterclockwise circle. In the expanse of the bay, this behavior looked ridiculous. It wasn't the tail lobbing that was odd, it was the diameter of the circle Keiko swam . . . almost precisely the size of the bay pen's confined north pool. Set in the sheer enormity of this new environment, the small course clearly illustrated Keiko's unfamiliarity with the new bay to even the most untrained eye.

Tom kept the session short, not wanting to lose Keiko by pushing his threshold for the unknown. After a few more quick behaviors and a marquee side breach, he pointed Keiko back into the bay pen. Asking Keiko to go back to a familiar place at just the right interval, Tom was making use of a learning process common in everyday life, rewarding the unknown with the known.

Media and bystanders left satisfied for the time being; however, sessions continued that afternoon repeating entry and exits to the bay. It was clear both from the protracted effort to get Keiko to the bay and his choice location between sessions, that he was not at all comfortable with the new accommodations. For this and many other reasons, not the least of which was our own discomfort with the yet unpracticed bay operations, the gateway was closed. Keiko spent this night and a few more inside the bay pen as usual. Three days later, I arrived back on the island and debriefed with Robin and Tom. In vivid detail they related Keiko's successes.

In the few short days following the debut, they had continued increasing exposure to the bay and had a few opportunities to reinforce Keiko for going to the bay of his own free will. Even so, it was abundantly obvious that the bay would require focused conditioning. Keiko's "mansion" was not yet the preferred locale that it was believed it would be.

The circumstances of Keiko's introduction to the bay created a frustration that boiled in my gut like a hot cauldron. I never spoke up; it was far too late for that, and I knew it would only generate friction between Robin and me. He would take the responsibility; he didn't shy from such things, but that wouldn't unshoot the gun. Still, it was not easy to let go of my disappointment. The forced introduction placed continued dependence on human direction, and at the same time it sacrificed the chance to capture self-motivated exploration from Keiko.

Every time I let it go, figuratively, I could only circle back to the missed opportunity and how rare it was in our quest for his freedom. Nevertheless, it was only March, and we were in the bay. That fact, independent of any other consideration, was worth a great deal.

Boy in a Bubble

Sadly, "I told you so" turned out to be the flavor of the first week of Keiko's access to the bay. We saw a complete setback from every expectation (or wishful desire) as a further set of complications reared its ugly head.

Keiko made no bones about his preference for the old familiar bay pen over that of his new playground. In the first few days of bay operations, he would only leave the pen at our behest. Given the freedom of choice, he would nest himself in the confines of the north pool, the innermost sanctuary of the bay pen. Never mind the open gateway and wide open expanse of the bay there for the taking. Repeating much of the structured conditioning of his first separation to the medical pool, we leveraged every possible tool at our disposal to encourage Keiko's voluntary exploration of his new digs, this time without forcing the issue.

Yet one more trait that challenged every notion of survival, Keiko was quick to shut down in the face of repeated failure. There was no doubt in my mind that the initial process of exposure to the bay had cost us dearly. At the heart of the matter, the process had demonstrated how Keiko adapted to change. We wanted a whale that would eagerly dive into new environments, chomp at the bit, seek out fresh and undiscovered territory and show extreme curiosity toward other living things (apart from humans). Instead, what we had was a withdrawn, neurotic introvert; dependent on our direction at every turn, void of even a spark of life at the onset of new challenge. Not what one would expect from a whale called "killer."

The best we could hope for was to set the stage, pique his interest in the bay, lie in wait and insure that nothing hindered each tentative step forward. Then, taking advantage where only a fingerhold existed, we provided familiar reward for each ventured and voluntary act that resembled the whale we envisioned for release. It was excruciatingly slow going at first. But the tenants of learning reign supreme. For each meticulous step forward, when all the elements are combined unfailingly, behavior will follow the path well laid.

By the third day of access, we were just beginning to see the fruits of our tedious watchful labor. Following sessions where we asked Keiko to the outside platform, he began to linger, whereas prior sessions had ended with his immediate return to the pen. Other times he would venture out of the pen on his own, making a brief appearance in the bay. The appearances were hardly worthy of report, but enough with which we could wield our trade: consequences that revived interest, shoring up his confidence and boosting repeat and prolonged performances.

Our rally was short-lived. On the fourth day, Keiko lost all interest in the world around him. Food held no value. In or out of his spartan accommodation, it didn't matter where or how it was offered. Trainers, his broken and distanced family, stirred nothing in the whale. He only sat motionless, the black of his melon camouflaged in the undercarriage of the pen's structural pipe. Keiko's only movement was facilitated by the undulating surface swell of the bay pen's interior. Something was very wrong, and this time it wasn't herring heavily laden in fat.

Just like illness in people, animals experience the same wax and wane of unpleasantness. In the breaks between whatever was taxing our athlete turned patient, we were able to get enough attention from Keiko to draw a blood sample. That afternoon, Robin ran the sample to the local hospital. It would be enough to get a basic read on that which we could not solve by observation alone. As we had expected, Keiko's blood work showed a spike in his white blood cell (WBC) count. Not a nominal spike, his WBC was elevated to concerning levels.

Pinpointing the exact cause of an illness via blood sample results is an art form. Sure there are usual suspects that typically indicate the more well-known ailments. But in most cases, the best one can hope for is a general idea of the nature of the affliction or at least clues enough to prescribe effective treatment. In this case, based on various key results in his blood panel, Robin and Dr. Cornell consulted by phone and determined that the condition was likely an infection. The culprit could have been in his stomach, his urinary tract or any number of skin or other commonly affected areas of his body. Nevertheless, given all the metrics at his disposal and the observations of Keiko himself, Robin believed the ailment was respiratory in nature. Although he was not a trained veterinarian, Robin had an uncanny Sherlock Holmes' ability at solving medical mysteries, particularly as they related to marine mammals.

Respiratory issues are not to be underestimated. Killer whales in particular are masters at masking illness. Their sheer size and substantial reserves make them formidable warriors, capable of sustaining serious illness over much greater time periods than humans can comprehend by comparison to our own fragile vessel. Even in the full-blown later stages of pneumonia—without routine clinical evidence—outward signs in a killer whale can easily elude the casual observer. But the condition is no less serious regardless of how obscure the symptoms. In fact, by the time the internal tidal wave of infection reaches a level where visual evidence presents itself, the condition is often so widespread that recovery is against the odds. Sadly, these are the hallmarks of many a stranded animal with which we had more encounters than we cared to recall.

In the face of a potentially life-threatening infection, priorities change rapidly. We didn't throw the book out the window as it related to Keiko's rehabilitation, but we also didn't attempt to make any progress. For the time being, the bay would have to wait. Our new goal was to get enough food to Keiko to prevent dehydration. We also used food to administer medication. Without delay, Keiko was put on a regimen of Tribrissen, a fairly focused antibiotic used to treat respiratory tract infections, and amoxicillin, a more moderate spectrum antibiotic that provided insurance against untold possible bacterial offenders. Our first hope was that we had caught the freight train before it gained too much steam.

During the following three weeks we nursed Keiko slowly back to health with guarded optimism. Though he responded initially to the dynamic duo of Tribrissen and amoxicillin, necessary adjustments in treatment became apparent as his WBC again elevated after weeks of initial positive response. Believing the condition was clearly related to a bacterial infection, a much more potent antibacterial called enrofloxacin was prescribed to replace the original cocktail. By March 24, 2000, nearly six weeks following the initial scare, we finally saw both clinical and outward signs that Keiko was out of danger. As would be expected, his treatment was maintained for a period beyond the proverbial finish line, well long enough to eradicate whatever mysterious villain had overwhelmed his immune system. Despite our delicate but growing distance from a weakened state, the Behavior Team was quickly back to advancing Keiko's acclimation to the bay.

By the end of March, although Keiko remained largely disinterested in food, he began to gain confidence within the expanses of the bay. As the progression into Phase II unfolded, it seemed no time at all, and we were back on the fast track in pursuing the formidable challenges in pursuit of release.

The Test

The operations team had been pining for a real test of the barrier net. Within the first weeks of Keiko's access to the bay, they got what they wanted, handily. It didn't take a pronounced storm front. Rather the customary winter weather of Klettsvik was all that was necessary to flush out the more notable aspects of the net's weaknesses. Unyielding winds and inconstant currents in early March took their toll, tag-teaming the barrier net from surface to bottom and in between. The varied exposure offered a gradual demonstration of the net's Achilles' heel and afforded Mighty Mo sporadic lulls in weather to make adjustments.

Over the course of three such days, the team anxiously inspected its creation, following a contrasting three days of assault from the elements. Damage unseen from the surface nonetheless abounded in the murky depths. Where the team had worked at intervals to shore up the weaker points of the net's anchoring, Mother Nature worked tirelessly to wear them away. Helicals that had been driven two feet into the dinosaur-aged bedrock were uprooted like frail weeds. Anchors on both sides of the net had been moved as if they were but toys. The "big ass" chain that laid straight the bottom length of the net was now serpentine in its course between the rock faces that framed Klettsvik Bay.

The operations team made their repairs, and fortified the weaker points of attachment that had been pinpointed by the fierce currents. Throughout its existence, despite redoubled efforts, maintenance of the barrier net was a constantly raging battle. On one side a frail human contraption hopelessly reliant on mending and cleaning. On the other, a relentless Klettsvik determined to throw off her involuntary muzzle.

Apart from dominating operational struggles, subtle changes in Keiko began to take form. Silently, in the shadow of Klettsvik's looming rock walls, Keiko turned a corner.

8

The Mean Season

Settling into April, operations became firmly rooted in the expansive bay. Now, more than ever, we focused laser-like attention to Keiko's choice of location. Though we often asked him out of the pen at our direction, Keiko's voluntary investigation of the bay was the prize we sought. At first these golden moments were scarce. But like baby steps, Keiko gradually gained interest through repeated trials and, by no small measure, through our efforts to encourage his exploration of the bay.

The monotony of watching sometimes hours for an opportunity to "catch" Keiko acting the part of a wild bull killer whale were no more. Now, we found ourselves hiding in packs around the bay pen, each of us interested in when he would come out again, where he would go, how far he might venture. We each wanted to be the one that might stretch his territory to new heights in space and time. Day by day, Keiko gave us more material with which to work. Eventually it became commonplace for him to travel so far from the pen that he appeared as if a small dolphin in the distance. The sight was intoxicating.

Ultimately Keiko adapted to the enormous bay more quickly than we did. At times, three of us would station ourselves on three distant corners of the bay pen, and yet it could take fifteen minutes or more to locate Keiko within the bay. Truthfully, our inability to supervise Keiko in the bay was the root of much discomfort on our part and therefore we continued to separate him back into the small area of the bay pen each night.

We overcame this insecurity once we recognized that Keiko's happenstance run-ins with the anchor lines were minimal and decreasing. The irony of our own fear of the unknown paralleling with that of Keiko's seemed a fitting reminder of the complex forces at work on both Keiko and the release team. As we finally relaxed into our role, Keiko began spending almost all of his available time in the relative vastness of the bay. He only came into the pen when called in or during heavy storms.

HDS

Delivering Keiko's reinforcement in the bay was a logistical challenge that required a creative solution. It also proved to be immensely enjoyable for the Behavior Team. Recall that at this stage, our principal tool for reinforcing or rewarding appropriate behavior was food, nearly 100 pounds of herring and capelin per day. But even the most amazing pitcher in the major leagues could throw a herring only so far. Many of us developed painful "pitcher's arm" in the pursuit of competitive fish-throwing. If we didn't find a way to deliver herring to any part of the bay, we would only teach Keiko to swim within an "arm's throw" of the bay pen.

We entertained ideas of CO_2 powered "fish cannons," stationing "pitchers" around the cliffs of Klettsvik and even seeding the environment with disguised herring-boxes throughout the bay. Most proved to be impractical either due to physics or operational limits. Ultimately, we decided on a favorite of many third-grade sharpshooters, a whale-sized version of the wrist rocket: a slingshot.

In no time, Mighty Mo had a local metallurgist fabricate what looked like a giant fork missing the center prongs. They attached several bands of rubber surgical tubing to each side of the slingshot frame and looped them through a funnel situated in the middle. We dubbed our proud creation the Herring Delivery System or HDS for short. We found the obvious attempt to disguise the toy in a formal title to be quite humorous. The HDS was affixed on top of the research shack, firmly rooted in a larger pipe bolted to the roof. This way we could pivot the slingshot in any direction. If the boyish backyard wars of childhood were carried through to an adult version, the HDS was the unbridled result, equivalent to a .50-caliber mounted machine gun.

It was more powerful than we anticipated. On its first use, the HDS nearly obliterated the funnel that acted as the sling. True of many prototype devices, we had to go through a bit of testing, and though we worked out the kinks pretty quickly, the HDS still required no small amount of skill. The operator had to lean back at a precise angle, gauge the wind and "hold his tongue just so" . . . but once released—*THWHAAAP*—the HDS would unleash a violent barrage of North Atlantic herring high into the air and well distanced. As we each became proficient, we could launch five or six individual herring simultaneously and up to 500 feet in almost any direction (further when we had the wind at our backs). We could also put the fish where we wanted with precision. As for Keiko, he now encountered "fish from heaven" seemingly at random and in all corners of the bay.

There were a few unanticipated benefits of the HDS. For starters, it made DRA conditioning fun and therefore increased the amount of time individual team members spent looking for appropriate behavior to shape. Also, the result of herring floating at the surface and spread in random fashion was not unlike the aftermath of wild whales stunning a biomass of herring with their flukes. Icelandic killer whales often forage this way. When observed from a boat, the feeding grounds of wild whales feasting on herring were left riddled with immobile fish scattered and floating at the surface.

The birds of Klettsvik quickly learned that the telltale *THWAP* of the HDS meant free food. When the herring hit the water near Keiko, sea gulls the size Christmas turkeys would swoop down and steal the fish right from under Keiko's nose. On a few rare occasions, the most talented gulls would grab the herring in mid flight at the peak of its trajectory. When the ricocheting winds of Klettsvik were at their most disorganized, the wild and random paths our flying fish could take never ceased to amaze us, their final destination no less confounding. Often this increased the level of difficulty for both whale and birds.

It didn't take any time for Keiko to learn there was competition for his food. Soon he was reacting to the herring much like a predator would react to an opportunistic prey. I was thrilled. As hard as we tried, I had never witnessed even the best of trainers produce predatory reflex reactions like those imposed by the live-or-die ultimatums demanded in nature. The HDS provided food for all that lived in the bay, and the winner was determined by who wanted it the most. That competition sometimes frustrated Keiko, but his apathy quickly turned to aggression directed at the occasional seabird. In the momentary pandemonium at the water's surface, there were more than a few chances for Keiko to add waterfowl to his menu, which he quite nearly did, once or twice ending up with tail feathers instead. The frenzy created by the HDS sharpened Keiko in ways that nothing else could.

FLASH REPORT

Staff Update, April 1, 2000

Latest information regarding Iceland operations: Keiko

Who'd thunk it! In an extremely spellbinding moment . . . Stephen Claussen was left speechless after Kelly "the Funnel Queen" Reed nailed a Great Northern Diver (a bird) with the HDS (Herring Delivery System) on live television. In a statement following the incident, Kelly eloquently side stepped the issue by drawing attention to Greg's accidental marriage to a local Icelander late last night.

So many wild whales have entered the area that DR with Keiko has become increasingly difficult. Yesterday alone, seven other whales took over 40 pounds of Keiko's base when he reacted too slowly to retrieve the fish. Steve Sinelli was seen later that evening shooing the wild whales away during a husbandry session.

Marine Operations has been missing for several days now. If anyone has any information . . . there is a

30,000 Kr. reward for information leading to their arrest.

The barrier net continues to spend more and more time in the bay while Keiko holds fast. As long as the weather remains calm we expect to have all the kelp cleaned off of Keiko this week.

Ocean Futures latest press release was extremely successful. Jeff Foster conducted the interviews which captured the community's attention with a colorful description of Keiko's attraction to the Draupnir.

Our main office instituted a new policy regarding residence at the hotel here on-site. Firearms of any kind will no longer be allowed in your rooms. Sorry, Blair.

We have recently had to reduce Keiko's exercises to one behavior only when in the bay. This was decided in a group discussion after a perimeter lob-tail that Steve sent took 32 minutes to complete.

April Fools

The Iceland Team

Boat Etiquette 101

Another sweeping change in Keiko's life during Phase II was to introduce his newly acquired high-seas escort, the *Draupnir*. I was beside myself with anticipation to get started on boat walk conditioning, training Keiko to "heel "alongside the *Draupnir*. Initially, we labored through tedious planning and communication separating Keiko into the bay pen when boats entered the bay. We were not sure yet how he would react. Still, it was an inevitable encounter that could not be avoided, so it had to be conditioned.

Through time and repeated exposure to these boats, making sure that they offered little in the way of interest, we knew he would eventually tire of them and move on to more interesting stimulation in his environment. This is how the process of desensitization works. When something becomes so mundane and offers no value, that object or stimulus slowly and surely fades from our awareness as if invisible. But this process takes time. It also requires some degree of counteracting the boat as a naturally appealing break in an otherwise monotonous bay. To do this, we would randomly provide reinforcement for the chance few times that Keiko ignored the boats, then build on that foundation.

The first few times we allowed Keiko to have access to a boat was when the *Sili* made her way to and from the bay pen. The close encounters were largely uneventful, although he shadowed the *Sili's* every movement from the curious crossing of the boat gate, to the pen, and back out again. Always right on her port or starboard beam, he would cock his head to one side and eyeball the occupants as if to say, *Hey there, what-cha doin'?* The staff was well instructed (warned . . . okay, maybe threatened) not to give one ounce of attention to this silent but humorous inquiry. So clear were the protocols of desensitizing Keiko's interest in boats that even eye contact was forbidden when he was soliciting attention from any boat, even the *Draupnir*. During the course of these first steps, we changed up the schedule and the boats, as well as where they went and how they went with great creativity. Much like training a police horse to ignore almost any unpredictable event that might occur in a crowded public park, we were exposing Keiko to as many versions of a boat as possible. We even recruited the occasional third-party boat in order to expose him to unfamiliar craft. In every case, beyond that of providing something to watch, he received no direct form of response or reinforcement from the waterborne citizens of Klettsvik.

The next step in the process of conditioning the walk-boat was to create something very distinctive about the *Draupnir* and when Keiko was allowed to approach her. It's important to note here that conditioning and desensitization are, in simple description, opposite forms of shaping behavior. Conditioning (also shaping or modifying behavior) is an active process and involves creating reinforcing consequences. Desensitization seeks to eliminate any form of reinforcement—for that matter, any form of stimulation—associated with the event or object. An object can also have distinctive characteristics that determine when it has value. The absence of those same distinguishing traits can likewise render the object useless or having no value. Sometimes we see it, sometimes we don't. A phone that sits on the wall is invisible to us, until it rings. We tune out innumerable things in our background, until something changes, something unusual in appearance or sound or a combination thereof that alters the item's meaning for us. Although a crude description, this basic premise marked how Keiko's walk-boat was set apart from other boats.

Transferring his familiar platform was the first and most obvious of the distinguishing factors that set the stage for *Draupnir's* special attributes as the walk-boat. Over the recent two months in the bay pen, and even on his first weeks in the bay, we introduced the platforms as Keiko's only means to receive human interaction. Simple in their design, these platforms were approximately two by four feet, consisting of a continuous aluminum tube about three inches in diameter forming the rectangle of the platform. Nylon straps were cross-woven through the length and width of the rectangle and formed the surface where the staff stood or sat. Light, hydrodynamic and impervious to the water, they were intended to be an easy adaptation to Keiko's tailored and peculiar walk-boat. The platform was attached, much as they had been on the bay pen itself, to the outside pontoon or the sponson of the boat. The long side of the trampoline-like fixture hinged against the *Draupnir's* starboard beam, the outside corners supported by guide ropes tied off to the top of the pilothouse. This allowed us to retract the platform to an upright stowed position or "present" the platform by putting it in a down position resting nearly flat on the water's surface.

Next she was equipped with a small underwater transmitter fabricated by Woods Hole Oceanographic. The transmitter was used to create a short burst signal set to a frequency well within Keiko's upper hearing range. The combination of the call tone followed by the deployment of the platform defined the conditions by which Keiko was allowed to approach the walk-boat. Outside of the bay pen, the tone recall and platform were limited to the *Draupnir* and the *Draupnir* only. Like the platform, we had introduced the call tone months prior to bay access and the kickoff of boat-walk training. Both variables were well-rehearsed.

Such was the setting on Keiko's maiden "steps" with his escort walk-boat. First he would learn to ignore the boat; then, by design, he was taught to approach her, and to eventually follow her faithfully. During the first approximation, we began by having the *Draupnir* within the compass of the bay, milling about at random and without consequence to Keiko. Starting with our morning session, we dropped the west-side platform on the outside of the pen, triggering the call tone just as the platform hit the water's surface. At Keiko's arrival, a routine body exam ensued while the *Draupnir* motored by, only thirty feet off. On her second pass, the *Draupnir* stopped at much the same distance, this time holding her position and presenting her starboard platform.

"Bay pen—*Draupnir,* copy?" Even amid the best of conditions within Klettsvik, radio communication was necessary. Distances, birds, and wind all stole away any attempt at oral instruction. This day it was the birds. The high decibel disharmony of chirps and whistles squelched any chance of conversation even within close quarters.

"Bay pen, go ahead, *Draupnir*," Tracy replied. She was standing just behind Brian who was working with Keiko on the pen's west platform.

"I'm going to drop the platform then let's go right to the A to B. Okay?"

Located on the *Draupnir*, I took personal responsibility for Keiko's walk conditioning. Some within the Behavior Team took offense at this, having had more "time in" with Keiko and the project. I was keenly aware of the sentiment, but also willingly sacrificed what equity I might have had with the more slighted team members. To me, the ever-increasing need for precision in the steps toward reintroduction was palpable. I wasn't about to take any chances with this important part of Keiko's training.

Brian waited until I had the *Draupnir's* platform in position. I sat ten meters directly opposite Brian, kneeling on the nylon trampoline suspended from the starboard sponson. As Brian stood and pointed directly toward the *Draupnir* with his arm fully extended, Tracy popped over the radio.

"Pointed."

"Copy," I replied, simultaneously hitting the call tone.

Keiko arrived in front of me as if he'd done this a thousand times, head cocked to one side and body trailing to aft in the slight current. He had never been quite this close to the boat. His length was nearly two-thirds that of the *Draupnir*. Michael had her in gear, idling to hold position, and leaned out of the pilothouse.

"He-hey, there he is!" Michael voiced the excitement we all shared.

I bridged Keiko and offered him a couple herring, which he took without pause.

"Bay pen—*Draupnir*. I'm going to point him back . . . let's repeat the same, then we'll start moving," I advised Tracy through the mini-shoulder mic on my radio. Brian and I exchanged Keiko in the same way one more time, to the pen platform and then back again to *Draupnir*.

Keeping my eyes on Keiko, I yelled to Michael who had now opened the side window of the wheelhouse.

"Okay, Michael, let's start at a very slow pace and see how it goes."

Michael throttled up, creeping the *Draupnir* forward. Keiko stayed in much the same position, head slightly aside and watching me on the platform. The movement required little to no effort on his part.

"Michael . . . a little faster."

The sound of *Draupnir's* twin engines pitched up as Michael applied slightly more throttle.

Keiko began swimming, now having to put his head down in a more natural position to keep pace beside the boat.

"How's that?" Michael tested. Three feet away, I struggled to figure out the best position on the platform, now awash with water. Standing and holding fast to the guide ropes was the only solution.

"Perfect . . . just hold right there . . . keep as wide as you can, no sharp turns," I said. The instruction was unnecessary. Michael and I had thoroughly vetted the plan earlier that morning.

Approaching our first lap around the bay pen, Keiko began drifting off position, going deep and slightly beneath the platform. Having a short target pole with me, I tapped the surface with the buoy just to the front and outside the platform. Keiko came to the surface swimming with his nose just above the water and touched the target. I blew my whistle and yelled at Michael. "All stop!"

I wanted to reinforce Keiko and needed to stop the boat so Keiko could sit up in his customary position at the platform. As the session and the day wore on, the process became much more fluid. In the first few laps around the bay pen, Keiko frequently drifted beneath the *Draupnir*, perhaps taking advantage of the hydrodynamic slipstream created by the hull. (Slipstream is an area of reduced air pressure and forward suction created by the hull when a boat is moving fast.)

At an early pit stop, Tracy joined me on the *Draupnir* to assist, handing me the occasional herring and/or the target pole. Anticipating and correcting Keiko's position, I prompted him with the target from time to time, but rather than stop the boat, Keiko and I both became proficient at the exchange. I tossed herring out to my left, speed dependent, two or three meters ahead of Keiko. He quickly learned to fast-grab the fish lest it disappear in both our wakes. Once or twice he missed the grab and turned back to retrieve the fish. This, of course, required the *Draupnir* to slow, allowing Keiko to rejoin the walk. As my aim at positioning the tossed fish became better, so too Keiko improved at snatching the morsel while keeping stride.

His best grabs were addictive. At higher speeds it was necessary to throw the fish at just the right angle and with force. Otherwise it would skip off the surface and well out of Keiko's path. But when I got it just right, not too close and not too far, Keiko would surface like a charging beast, water gushing from both sides of his open mouth as he snapped down on the fish. Once or twice I threw the fish in rapid succession, and although I made every attempt to put the herring in exactly the right spot, more than a few missed the mark. Astonishingly, Keiko quickly became skilled at moving his head to one side or the other to compensate, almost as if he had a neck. The "game" seemed motivating for Keiko. Watching him grab and dive as he did was certainly exhilarating for us.

By the afternoon, we had completed nearly two nautical miles in the first round practice walks. Fun for both the crew and Keiko, we all rapidly became adept at the rhythms of the slow dance. Yet the level of communication and skill at handling the *Draupnir* was vital. Only Michael or Greg would captain the walk-boat during these pivotal rehearsals.

Within a few short weeks, we were routinely practicing boat-follow training and increasing the distance Keiko traveled in company with the *Draupnir*. At first, we set up only two or three short walks each day. We practiced not only the walk itself, but we also taught Keiko when to leave the *Draupnir*. At random intervals we stopped the walk, retracted the platform and went into a neutral position. The first few of these breaks Keiko remained at the side of the *Draupnir*. He peered up at her inhabitants and waited patiently. When he finally lost interest and moved away, the team on the bay pen threw herring. To the best of their ability, they aimed for Keiko's path away from the *Draupnir*.

Initially, the need to directly reinforce Keiko for leaving the *Draupnir* in her neutral position was frequent. But the practice became much more random and intermittent as rehearsals wore on. Eventually Keiko came to understand the clear signs of the *Draupnir's* distinctive walk stance. He also began to recognize when it was time to explore away from the boat. Platform down and recall tone meant "let's go." Platform up and a deck void of humans meant "go play." It wasn't long before Keiko became an expert at discerning the difference. An added benefit, his learning with the *Draupnir* in walk-mode helped to further desensitize his interest in other boats. By the end of the month, Keiko was no longer shadowing the random passage of waterborne vessels in and out of the bay.

It seemed no time at all, and the first baby steps had morphed into full-blown April walks. As the walk rehearsals increased in distance, they decreased in frequency. In time it became fairly routine to do only two walks per day; weather permitting, one in the morning and one in the afternoon. At our peak, we were knocking down over fifteen nautical miles a day comprised of circles and figure eights inside the bay.

Not bad, but still nothing compared to the travel Keiko would have to endure in company with wild whales. Traveling alongside the *Draupnir* during walk sessions was not, however, Keiko's only means of exercise. After more than a month in the bay, we had successfully pushed his activity level to extremes.

The use of DRA conditioning techniques was not unique by any measure, but the sheer intensity with which we applied DRA conditioning in this setting was indeed rare. Nearly ninety percent of Keiko's food was delivered covertly, away from human contact, and in response to active swimming. Any form of activity well-away from the bay pen and the presence of man-made items (i.e., boats, docks, the barrier net and the pen itself) were subject to random reinforcement. In addition, we were now watching for opportunities to encourage activity almost twenty-four hours a day, seven days a week. There was no predictable time off in Keiko's new world.

Bump in the Night

Late April 2000: Daylight now extended from five a.m. through ten p.m. Our opportunities to conduct walks or other forms of conditioning at almost any time had likewise expanded. Now proficient at walk rehearsals, Keiko was routinely circling and weaving alongside the *Draupnir* an average of ten nautical miles each day, most times in one continuous trek. Longer walks now made the bay seem confining. Going in circles for eight or ten miles was uninspiring to say the least. That being said, it was nonetheless vital to ensure that Keiko could sustain extended distances and at varying speeds.

During one idle afternoon, we stopped the *Draupnir* and assumed a neutral position, permitting Keiko to venture away from the boat. Greg was captain; Stephen Claussen and I were working with Keiko. By this time it was common to have two of the Behavior Team members on the boat. I worked with Keiko directly from the platform. Stephen assisted by managing the target pole and tossing Keiko the occasional herring. He would also periodically stand by our second HDS, mounted on the bow of the *Draupnir*, giving us a much improved range for providing reinforcement when Keiko explored the bay. At each stop, the platform was pulled up and tied off, communicating the "closed for business" stature of the *Draupnir*. Any crew on the boat immediately retreated to the interior of the pilothouse, thus completing the distinction.

In the close quarters of the cabin, the three of us exchanged small talk. Mostly we listened and watched for Keiko's usual departure to some other portion of the bay. But this time he didn't leave right away. Keiko remained just off the starboard side of the boat looking for any signs of life. Finally he moved off but returned just as quickly on the opposite beam, again lifting one eye above the surface toward the empty deck of the walk-boat. We continued to ignore the solicitation, expecting Keiko to go about his way as he had done so many times before. Stephen and I entertained each other with our usual witty banter. Then, without warning, the *Draupnir* swiftly lifted from the water. Instinctively, we each grabbed hold of various fixtures in the cabin to steady ourselves.

"Holy shit!" Greg reacted. "Did he just hit the boat?"

It felt as if the *Draupnir* had lifted a foot from the waterline and was held there floating. In reality it was mere inches and just a fleeting moment. As the boat settled back to her footing, a loud *thud* sent a shudder throughout the entire vessel providing exclamation to the surprise levitation.

We began shifting about the pilothouse and crowding around the small windows to get a glimpse of what was going on, hoping to locate Keiko.

Before any of us could offer a response, a second hit rocked the boat. This time it was much sharper; a blunt force *crack*.

"That's not good," Greg stated to no one in particular. Now he looked very concerned. "Should I move?" It was a determined suggestion more than a question.

"No, let's just wait a minute," I didn't want to react to this new behavior. Instinctively I knew that a reaction would only entice Keiko to continue or worse, increase the intensity of his assault on the boat.

"It sounds like he's going to crack the hull," Stephen said dryly, not at all helping to calm Greg's concern for the *Draupnir*.

Again, another *CRACK* followed by the momentary rise of the boat.

"Ohmygawd!" Stephen bellowed. "That was hard!"

This time Greg was beside himself. He wanted to take action. "I don't know how much of this she can take. We should move or call him over . . . we can't just let him keep hitting the boat."

"We have to wait. We're only going to make it worse if we react. We need to wait until he leaves the vicinity for at least a minute or so," I repeated.

No one was on the pen this time, so we had no way to get a bird's-eye perspective of what Keiko was doing. Even worse, we couldn't see Keiko from within the pilothouse unless he was away from the side of the boat. So far, he had largely stayed underneath the *Draupnir*, only passing from side to side as he struck the hull. From the unnatural movement of the boat, we could tell that he was also remaining beneath, rubbing or pushing his back up against the hull.

"Do you think he's doing that with his head or is it like his flukes or something?" Stephen was testing the various theories we were all visualizing. As if just occurring to him, he added, "He can't tip the boat can he?"

"I have no idea." And I didn't. "I'm guessing it's his head, although I can't imagine how he can hit it that hard without cracking his skull."

The thought given voice did not help Greg. His concern for the integrity of the hull now convinced me that damage was likely if not certain.

Draupnir rose again as if on a small swell, followed by yet another thundering *CRACK!*

"Jeez!" Stephen exclaimed. "Five feet. I think that was about five, no, maybe eight feet. Yeah, about eight feet out of the water."

"Thanks, Stephen, that helps," I retorted.

"Okay, that's enough . . . we've got to do something," Greg urged.

I knew Greg's patience was just about gone. Regardless, I was just as stubbornly holding fast to providing no reaction to whatever this was. As I strained to find Keiko out the window of the pilothouse I tried to convince Greg to stay put for just a few more minutes.

"Just hang on a second. As soon as he moves away we'll head to the pen." Crammed inside the pilothouse I was keenly aware of the stagnant smell of grease and maybe a little mildew. Turning to look out the opposite window I almost stumbled over Stephen, who was on his butt, knees pulled up to his chest and rocking back and forth as he pretended to suck his thumb. I couldn't help myself; involuntarily I let a short chuckle escape.

"Guys," Greg's one-word warning was stern enough.

"Stephen you're killing me here," I said trying to realign with a serious atmosphere. "At the least let's make sure we know where he is before we do anything."

In a perfect world, I wanted to wait until Keiko left the area and not just by a little. I wanted to be convinced he had lost interest in the assault and moved on to other things. As it was, I couldn't afford to wait that long or be that certain. Clearly Greg wasn't willing to chance it much longer. Thankfully, Keiko finally gave us the small window we needed. Only a short time had passed. At the time it seemed like an hour. Keiko moved away from the *Draupnir* and was, for the moment, swimming directly away from our position.

"Okay, let's get to the pen. Drop Stephen and me off, and then you can take the boat in," I said.

Greg quickly engaged the engines and moved to the west side of the pen. Responding to her movement, Keiko turned and headed back toward the walk-boat. He swam slowly at the surface in no apparent hurry. Greg had the *Draupnir* in reverse almost as quickly as her bow touched the pen, allowing just enough contact for Stephen and I to leap onto the pen. Although Keiko stalked the *Draupnir* on her exit from the bay, he did not strike her again that morning.

In the following days, Keiko repeated the strange behavior unpredictably, but almost exclusively during walk rehearsals when the *Draupnir* assumed her neutral position. He seldom hit the boat when she was transporting crew to or from the bay pen before walks began and not every time during walks. I began calling the behavior "love taps" in an effort to diffuse the heightened concern. Following the initial event and in a staff update, I made light of the situation, grossly exaggerating Greg's concerns for humorous effect. Though Stephen and I shared a bellyful of laughs in drafting the update, Greg didn't appreciate the embellishment in the slightest. Of course, this only served to fuel a more colorful tale at each telling.

The Wrong Way

Thus far I had been successful in preventing any direct or immediate reaction to the hits. But the need to fix the issue eventually led to an all-hands staff meeting to examine the problem and figure out how to address it. We congregated in the hotel solarium, our favored spot for any form of communal exchange. It was late enough that the starry sky was visible through the glass panels transforming the room into a nightly planetarium of sorts. Every soul on-site gathered to one side of the solarium, some in the few chairs that scattered the room, most on the floor or leaning against the half-wall leading down to the kitchen. As the discussion wore on, more and more creative ideas were offered in hopes of redirecting or stopping Keiko's beating of the *Draupnir*.

My frustration was blatantly obvious early on. Beyond describing the hits and the circumstances when it most often occurred, I largely listened to the speculation on potential damage to the *Draupnir* or Keiko self-inflicting injuries. There was no shortage of wild and some rational solutions.

Greg led the volley of alternatives. "Why don't we just keep the boat moving instead of sitting idle? Maybe that will keep him from hitting the boat."

Charles approached the subject from a logical point of view, seeking to break down the events. By now, he was becoming practiced in behavior analysis he had participated in with Robin and me. "Why do you think he's doing it in the first place?

"I have ideas, but the main thing is that we don't react to him hitting the boat once it's started," I couldn't disguise my impatience. I had already said as much in response to some of the more colorful solutions voiced earlier.

"If he keeps hitting the boat that hard, he's going to either crack the hull or hurt himself or both," Greg pressed.

"I understand. I'm not suggesting that I don't care about the boat or Keiko. I'm simply saying that changing what we're doing in response is only going to make matters worse," I said.

"It's like he thinks it's a toy . . . like he wants to play," Stephen suggested.

Again I didn't respond. I was tired and describing the "why" behind the odd behavior was completely useless at this point. No matter what the reason, it had happened, and we couldn't start guessing what might be going through Keiko's head. I knew that was a natural reaction but also a dangerous one. In my world, events that prompted the behavior and the immediate consequences that increased it were the pivotal points of any solution. That and focusing on what we wanted Keiko to do *instead of* hit the boat.

"Look, at this point we know he's most likely to start hitting the boat when we're neutral and usually during the middle of a walk session. He's only done it once or twice outside of that, and even then they were halfhearted bumps, not nearly the intensity as during walks."

I was trying to explain what we could predict rather than guessing at motive. "I've seen similar behavior before, although the outlet is different. It's basically frustration."

"Frustration from what?" Jen asked.

"Schedule-induced frustration," I answered. "We've been doing the walks now for almost two months. We've eliminated almost every other form of stimulation that he's used to." This all seemed so obvious, "The only time he gets 'us' is during the walks."

"So you're saying we need to go back to more other types of interaction?" Charles was probing for next steps.

Stephen jumped in, "He used to have a blue raft in Newport. He liked to push it up and knock us off of it. It was a game. Maybe he thinks it's like the blue raft."

Again the guesswork. My weariness got the best of me.

"Guys, I'm telling you this is frustration. We can't react to him hitting the boat no matter what the reason. If we do, he'll just start hitting the boat anytime he wants our attention." I couldn't resist and went *there*. "I've worked with over twenty-six other killer whales. I'm telling you I know what I'm talking about. I've seen the exact same thing before, it's frustration. We have to stick to the consistency and look for more opportunities when he doesn't hit the boat, especially during the times we've gone neutral."

I knew when I said it that I'd crossed an unspoken line. One cannot force agreement by imposing the "experience card." I had gone out of my way throughout the entire project, often in turbulent interactions with the more inexperienced staff, to avoid pulling the holier-than-thou punch. Until then. Stupid. I regretted it as soon as the words left my mouth.

Robin knew I was right. I knew I was right. For the love of God I couldn't figure out who couldn't understand this basic concept. If I had been surrounded by senior trainers from the SeaWorld of Florida park, these topics would never even enter the fray. It was exhausting and seemed such a baseless waste of time.

What we needed was to be sure there would be no knee-jerk reaction out on the water. This was critical and was foremost in my mind. Supporting that, we needed to monitor how we were reinforcing the "right" behavior of leaving the *Draupnir*, when Keiko was swimming away or exploring throughout the bay. One of a trainer's favorite cliché phrases, "rehearse is worse," beautifully describes the proactive need to anticipate conditions that lead to misbehavior and avoid allowing it to occur in the first place. We also needed to identify any behavior other than hitting the boat or even those incompatible with hitting the boat and invest our time and attentions there.

Greg wasn't satisfied. Probably most that night weren't, and they shouldn't have been. But I had convinced Charles that I could handle the situation. By now, he was willing to put his wager on my approach, even if he didn't understand completely, so poor was my tact in debating the topic in the open forum. Even though my stubborn insistence was fundamentally sound, I wasn't completely without fault in my earlier dismissal of Stephen's fleeting reference to the dingy used for play in Oregon.

What I did not know that night, I soon learned through ongoing review with Stephen, Tracy and Brian. The blue raft in Newport was more central to the behavior than I had realized. Although this discovery was relevant, it didn't change our approach. It just meant that it would be harder to extinguish than I anticipated. Keiko wasn't just hitting the boat out of frustration born of predictability in his walks; it was a long-standing behavior that had been part of a game in his Oregon facility. The object of the game for Keiko was to knock his trainers off the raft, often completely capsizing the boat-like toy.

When an existing behavior, such as knocking the blue raft, or in this case the *Draupnir* (guilty by generalization), is directly rewarded, and rewarded in a variety of ways, that behavior doesn't just go away when it is no longer reinforced. Before it completely goes away, it increases in frequency and intensity. This effect is called "extinction burst." I did not offer this information to many of the staff. I knew it would only reignite the fires of hasty quick-fix solutions, all of which would actually lead to the exact opposite result. Determined to see it through, I knew I needed to be glued to every walk session and diligently work to countercondition behavior other than hitting the boat as deftly as I could possibly manage.

Unwanted Behavior

Under the dictates of release, much of what defined Keiko during his eighteen years with humans would have to be forgotten or replaced. As the project wore on, the use of DRA became a fundamental cornerstone to achieving the impossible. This translated to presenting reinforcement for almost any behavior that was "other than" or incompatible with Keiko's sedentary behavior or seeking human attention.

The immense sphere of influence over Keiko's behavior grew each day and with it the team's responsibility to govern those influences seamlessly. Life in the bay represented the first "dress rehearsal" for transition to the open ocean. Everything we did in Phase II was setting the stage to prepare Keiko for the grand step. We had to think about not only what types of behavior we wanted to encourage and what we wanted to reduce, but also every possible persuasion that Keiko was subjected to at any given time.

The results were remarkable. After a relatively short period focusing on the formidable principles of learning, Keiko began showing signs of becoming an independent animal. He also literally never stopped swimming. There were times that we were unsure whether or not this was entirely healthy. Unless a storm drove him to float on the leeward side of the bay pen where he was shielded from the current, he never stopped moving. No doubt this activity level certainly pushed his stamina to new heights, but it could also introduce harmful stress.

We expanded our criterion, looking for inventiveness displayed by Keiko; things beyond simply moving. This could be rubbing on rocks that crept out from the cliff's footing or the inquisitive following of nervous birds on the water's surface. Provided that Keiko was not seeking us or watching the passage of boats and, of course, that he was not altogether sedentary, almost anything could be targeted by the HDS cannon.

Soon Keiko showed little to no interest in our whereabouts. He seldom paid heed to boats in or out of the bay. He often disappeared for five minutes or more beneath the surface close to the eastern rock walls. We never discovered exactly what he was doing, but whatever it was, it wasn't seeking human attention. Remarkable as the results were, the changes in Keiko were still far removed from a whale capable of survival on his own. It was one thing to produce these changes in Keiko within the context and confines of the bay and entirely another to transfer the new Keiko to a completely foreign world, one without the benefit of our involvement.

Throughout Keiko's life he was taught to follow. In another time and place he might constitute the world's best employee, ever dependable in following precise direction. However, we needed Keiko to learn to take initiative. More, we needed that initiative to become a permanent change in his life. The latter part of this equation speaks to the importance of "reinforcement schedules."

Schedules are the key ingredient in producing lasting changes in behavior. The interval, ratio, and variability of reinforcement can produce sustainable change, create unwanted dependency or completely eradicate a particular behavior. Like many advanced areas of science, the particulars of this process are enough to give one a nosebleed. Suffice it to say, in the later stages of Keiko's preparation for a life of independence, the careful management of how and when he received reinforcement consumed our every thought. If the independence we were seeking to shape didn't transfer to his new world, Keiko wouldn't have the slimmest chance of survival.

Imagine a child learning to keep his room clean. Mom or Dad can offer the child a reward each time he cleans his room. However, stopping at this step in the process only makes the child expect a reward each time he cleans his room. As soon as that reward is no longer offered, he quickly reverts to sloppiness. In order to establish the act of cleaning his room as a self-sustained change, the circumstance of having a clean room must become reinforcing "in and of itself." Initially, that child is rewarded each time he cleans his room. Next the rewards become variable and intermittent, but each reward is preceded with positive activities in the clean room. Over time, the "clean room" itself becomes the reward, and the child "feels good" about having a clean room. The state of having a clean room is intrinsically rewarding, and the behavior of keeping it clean becomes a long-term change in the child.

In Keiko's case, the "clean room" was staying in the open ocean with or near wild whales. We needed to make this environment "intrinsically" reinforcing for Keiko. If only it were that simple. Bound by the ethical responsibility not to deliberately interfere with or influence the wild whales, we could not directly reinforce the whales or Keiko when in their presence. However, we could "clear the mechanism." We could minimize the value of Klettsvik Bay in Keiko's world and create a desire for stimulation that we hoped might be fulfilled by his wild cousins.

We prepared for this as early as May 1999, when we began reducing the amount and variety of stimulation in Keiko's life inside the bay. In effect, we systematically made his day-to-day world in the care of man sustaining, but boring. In fact, it was downright solitary confinement. Certainly the transition from the bay pen to the expanse of the bay initially offered Keiko some rather interesting adventures. But eventually the novelty of the bay was lost, and it became just as static as any form of containment where variety and stimulation are no longer supplemented by his human caregivers. This absence of variety in Keiko's life effectively created an enormous hole that theoretically "wanted" to be filled.

I always referred to the months approaching summer as "the mean season." By design, it was a period empty of the traditional things Keiko was lifelong dependent upon for daily stimulation.

This period was difficult for the staff as well, so long accustomed to varied and playful interactions with Keiko. Distancing ourselves from that which we adored, for the sake of his survival, though understood, ran afoul of our natural inclinations. For some, whose understanding of the process fell short, it was downright mean. As the staff struggled with letting go, Keiko himself became indifferent.

Innocence Lost

If finalizing the permit for release was the only outstanding item required to begin voyages to the open ocean, life would have been grand. As it was, multiple pieces of the puzzle remained unfinished. We were not yet ready to leave the bay halfway house. Boats were not fully equipped, at least not to the standards Michael required, and no one would question Michael's tedious demands on maritime preparedness. Bay pen and barrier net maintenance faithfully consumed their share of time, and of course the challenges of managing Keiko's needs and navigating through conflicting activity surrounding him were never-ending tasks.

Nevertheless, we had done our job well . . . too well. Keiko moved ceaselessly. In fact, at times it seemed the only thing that might stop him was a heart attack. Again we adjusted our focus, but the delicate balance teetered from one aberrant behavior to the next. It was like trying to steer a jet at Mach II with only a rearview mirror.

In the middle of our efforts to navigate this deprived state, one thing was clear; Keiko's inward struggle was just beginning. His world was upside down. Those things that had always yielded human affection and attention no longer produced the expected result. Everything he had known was now elusive at best, at worst, absent any form of acknowledgment. What feedback could be resolved in his new world was detached and uncertain. The bay itself offered little to the benefit of living things. No warmth, no acceptance, not even recognition. The distance between Keiko and the world around him was palpable. We had sought to create a void. That much we did, and the success of it was hard to endure.

As Icelandic daylight steadily stretched to its zenith, the world's most famous whale began to transform. In the staff's relatively short time with Keiko, he had been known for his playful and affectionate nature. Here, after less than three months into Phase II, neither his disposition nor his physical appearance represented the Keiko they once knew. His morbidly obese body became lean and muscular. The characteristic floppy dorsal fin, which previously laid over and rested against his body now twisted into a half corkscrew pulled upward and away by the constancy of his movement. His friendly curiosity in anything human was replaced with cold disinterest.

So stark in contrast was this whale to the famous Keiko, the change was apparent even to those who had never met him. On one such occasion, my wife, Alyssa, traveled to Vestmannaeyjar. Nearing summer and the prospect of open ocean exposure, my rotations on-site were increasing, and the only way we could see each other was for her to come to me. Less than thirty-six hours before her arrival on the island she had been working with eight other killer whales at SeaWorld of Florida. I was looking forward to her contrasting assessment.

Eye of the Tiger

Tilikum was over twenty-two feet in length and weighed in at nearly 11,000 pounds. Like Keiko, he had come from Icelandic waters. Collected around the same time the two males were also very close in age. Unlike Keiko, Tilikum had been in the company of other killer whales throughout his entire life with man. Further delineating the two, Tilikum had killed before. Subjected to subpar training methods and unrehearsed in the etiquette of whale and human water interaction, Tilikum had drowned a trainer in his pool at Sealand of the Pacific in Victoria, British Columbia. Barely a year after the death he was acquired by SeaWorld and moved to its Orlando park. Alyssa worked with the infamous whale daily.

Known as "Tili" by his immediate trainers, he was a very dangerous animal. Not because he was inherently aggressive by nature, but because he was exactly the opposite. Much like Keiko, Tili was approachable. In fact it could be said his disposition was friendly. But knowing his history, no one judged this whale by his cover. Perhaps most frightening was the fact that Tili had never learned how to treat humans in the water, something taught to almost every killer whale in the care of man at a very early age. This missing link, a haunting attribute for any killer whale, was multiplied by his sheer size, notably larger than Keiko. But this alone was not the only ill-fated trait that lent to a menacing state of affairs regarding Tilikum.

In his adolescent years, his trainers had inadvertently mishandled the use of toys in his environment, namely their retrieval. As a result, Tili became violently possessive of any object that entered his domain. Tilikum's abrupt Jekyll and Hyde posture apparent when he seized the occasional barrel or rope toy sent chills down the spine.

In other ways Tilikum was very much a typical bull killer whale. He learned quickly, displayed the usual overzealous interest in females and was remiss of intimidation. The only element in Tilikum's world that gave him pause was the dominant alpha female in the social group of whales.

These characteristics defined an animal that required tireless concentration from those that worked with and around him. This was the bull killer against which Alyssa would involuntarily measure Keiko.

She arrived to the island of Heimaey in the afternoon. Having made the trek from Keflavik to the small local airport in Reykjavik and on this one fortunate occasion, weather on her side, she was spared the wild ferry ride in favor of the commuter flight. Expecting to transition into open ocean work soon, we had held back the full complement of staff. Our ranks were unusually thin. Nonetheless, I managed to get off the pen in time to pick her up at the island airport.

Alyssa is always smiling. More than just a mouthy grin, she smiles with her eyes, in fact, with her whole being. This brilliant expression always warms my soul. It was her smile that had me spellbound when we first met and the same one with which she now greeted me on Heimaey. Clad in her winter apparel, her long dark hair pouring down over her turtleneck sweater, I was restored by the sight of her as she walked across the tarmac from the small plane.

Alyssa is a striking woman. A quality of her Eastern-Bloc ancestry, she is a strong woman. She can handle herself alongside any man of like size. Just as easily she can waltz with elegance at a formal banquet; the mannerisms and bearing of a true lady. I had far exceeded my station in wedding such a partner. Stunning by any measure, she did not lack choice. Yet I was ever confident in the union. Defying any disproportion in physical appeal, it was our friendship that bound the two of us. In our past we had proven to be a powerful team, she, the finisher and I, the starter. We complemented each other in both life and work. I found myself wishing more than anything that she could be at my side deciphering each challenge and fulfilling the epic task at hand.

Avoiding the mundane topics of our domestic lives, we spent the evening on a brisk tour of the island. We then carried on sharing our fascinations of the northern land through dinner at Lanterna, the finer of the limited few dining experiences. We kept the evening short in anticipation of an early start to the following day and what would be her first experience "walking" a killer whale alongside a boat.

The next morning I went to the bay pen at daybreak with the opening crew. Alyssa remained asleep at the hotel. On her clock, it was only just midnight. During this phase, the Behavior Team was continuing desensitization work with auxiliary boats, which required no small amount of communication between the vessels navigating carefully within the bay. The objective was to simulate the formation that would carry us to sea. Alyssa listened intently as the chatter choreographing the first walk-session of the day emanated from the base monitor and echoed throughout the hotel's concrete walls and tile floor.

At sea, the *Draupnir* and Keiko would always be shadowed by a support vessel. This was important for myriad reasons, not the least of which was to protect the nucleus of the walk from curious third-party boats. In rehearsals, we frequently assumed the neutral posture with the *Draupnir*, encouraging Keiko to leave the walk-boat and explore. Here in the bay, where we had control of the mock boat traffic, we could ensure that Keiko would not receive disruptive attention from the assortment of watercraft. With each passing week, we continued to introduce additional variables, eventually even attempting to draw Keiko's attention away from the *Draupnir* during walks or when he was free swimming in the bay. Careful in our estimations and each successive step, Keiko began to understand that boats, aside from the distinctive walk-posture of the *Draupnir*, were nothing more than backdrop.

By midmorning, Greg and the *Sili* fetched Alyssa from the harbor and brought her out to the bay pen. Cheating every e-mailed description of the harsh conditions in Iceland, and particularly Klettsvik, the day was flat calm. Rock edifices surrounding the bay were cast in bright sunlight, crisp edges and hard shadows exaggerated the angular surfaces. Vibrant green mountaintops were full of a congregation of birds taking roost and chattering softly. Klettsvik was in rare form.

After a short tour of the skeletal man-made pen, she joined me on the *Draupnir* for a continuation of the morning's practice walks. Alyssa wore a borrowed Mustang survival suit. It was the polar opposite of her usual close-fitting wetsuit I had been accustomed to seeing her wear at SeaWorld. She was lost in the Pillsbury lumpiness shouted in bright orange, but cute nonetheless.

"Couple things. . . . " I had waited until after we boarded the *Draupnir*. "If he approaches the boat or even when we've called him to the platform, we don't give him any type of attention, not even eye contact. Only the person designated to work him from the walk platform has any direct interaction, even then only when the platform is out."

Alyssa didn't require a lengthy explanation. She was already familiar with the conditioning plan that I constantly bounced off her through e-mail and phone conversations. Still, I was careful to treat her no differently than the rare guest who accompanied the walk-boat from time to time. Alyssa confirmed her understanding with an eager smile and nod. She had her hands stuffed in the pockets of the Mustang suit, arms locked straight and her shoulders pulled up around her neck. She looked as if at any second she might start twisting side-to-side and batting her eyelids. "Where do you want me to be?"

This time Michael responded. "You can stay in the pilothouse with me; if you want you can stand just inside the aft doorway. You might get a better look from there." His tone and posture carried an air of "preferred guest" about them. I was greatly appreciative of Michael's welcoming nature toward Alyssa. The few previous guests of the *Draupnir* during walks were there by mandate, members of the board or grace-and-favor appointments for local government or media. Even those were few and far between. We were very shielding of attention surrounding Keiko, especially at this late stage in the release.

"*Draupnir*—Bay Pen/*Heppin*." I radioed to the inhabitants of the pen and to Greg onboard the remaining support boat. In the rare calmness of the bay, *Sili* was tied up to the pen and would sit this one out.

"Go ahead, *Draupnir*," Tracy responded. Greg double-tapped the mic, acknowledging that he was listening.

"We're going to move out to the west side and call him directly to the *Draupnir*. We'll do a few laps and then go neutral." Tracy needed to stand by the HDS on the roof of the research shack. Following the earlier routine of the day, Greg would motor about the bay opposite the *Draupnir's* location. I continued, "This will be a fairly short one. Probably only two or three nautical miles."

Michael maneuvered to position and held steady.

"*Draupnir*—Bay Pen. Do you see him?" I asked. It was vital to know where Keiko was and what he was doing before we introduced direct interaction. We had to make sure that Keiko wasn't following the *Heppin*, sitting idle or watching human activity in any way. Waiting, I placed the portable recall speaker into the water by draping it over the gunwale and lowering it about a meter deep in the water column. It was a small device, roughly the size of a film canister, only twice as long.

A few minutes later, Tracy responded from the bay pen, "He's in the corner, Zone 1, near the boat gate. Maybe rubbing on the rocks there."

Even in the best of conditions, it was often difficult to find the twenty-one-foot whale that was dwarfed in the relative vastness of the bay. The sighting could easily be missed unless we were looking in the right direction at the brief interval when he surfaced to breathe.

I dropped the platform and keyed my handheld radio as I stepped over the sponson and onto the nylon mesh. "Recall." At 210 pounds, my presence on the extended appendage caused the *Draupnir* to list to starboard. Saltwater covered my bright orange boots to the ankle.

We didn't know how far away Keiko could hear the underwater tone. Just the same he never failed to respond. It seemed forever before he showed up, approaching from the port side of the *Draupnir* at depth. A somewhat eerie sight, his presence was first evident from the glowing white of his eye patch, disembodied from the rest of him by the turbid water just a couple meters below the surface; a rare vision granted only by the extremely glassy surface of the water.

"Okay, Michael, ready when you are," I said.

The unmistakable *thunk* of the engines engaging was the only confirmation required. The *Draupnir* moved ahead, immediately assuming the practiced pace of three to four knots. Keiko, hearing the *Draupnir* drop into gear anticipated the movement and was already head-down and moving alongside. Alyssa, who had been conversing with Michael, watched from the cabin door. Now silent, I knew without looking that she was watching intently. Although she had kayaked with the killer whales at SeaWorld, this was nothing of the sort. Set within the broad stretches of the bay, it was as close to the open ocean as one could get without actually being there. Knowing Alyssa as I did, I experienced the novelty vicariously, imagining what she must be thinking of the surreal scene.

As we rounded the southern end of the bay, we increased speed enough to force Keiko into a mild porpoise, akin to a trot. At this speed, he broke the surface high and with purpose for each required breath. In between, he swam in a much more hydrodynamic position a meter or more beneath the water's surface. At times, Keiko would turn on his side, the undulation of his body accentuated by the glowing disruptive coloration moving back and forth, propelling himself in concert with the *Draupnir*.

"Michael, let's take it down a notch for a bit," I said. We had been at the faster clip for almost two rotations around the pen, roughly one nautical mile.

As the *Draupnir* slowed, so too did Keiko, taking a series of breaths at her side. Alyssa stood fixated, edging slightly outside the doorway for a better look. While she was peering over the side, Keiko surfaced. Rather than taking a quick breath and going back to a head-down position, this time he kept his enormous head above the waterline and for a few brief heartbeats, the deep black of his eye cast a penetrating lock on Alyssa. It was a moment that neither she nor I would forget.

Later that day over the customary glass of red wine, Alyssa shared her impressions. Her view of Keiko had been colored by my doubting e-mails and descriptions. Yet what she saw in that one day, her singular experience, deceived every portrayal she had heard of Keiko. Alyssa described his gaze as a cold indifference, if not the menacing stare of a killer. But it was more than that, his business-only interaction throughout the walk and subsequent disinterest in the trivial goings-on of humans around him whispered hints of the same royal disposition she had witnessed from wild orca in the Pacific Northwest. The changes in Keiko had been gradual. I had been too close to adequately gauge them.

Alyssa's blunt observations of a point in time instantly revealed how far we had actually come. What she described then, the unforgettable gaze of a whale I once believed least like any killer whale I had ever known, was now compared to a wild whale. If I had made the offer, she would not have gotten into the water with the animal, such were her misgivings. This, her trusted measure of Keiko's temperament, was no small victory. It was indeed everything. Even I, at times certain we were marching to war already defeated, began to believe there might be a very real chance for the Big Man.

Seldom able to revel in what progress we could measure, the nature of the release effort meant we were constantly unhinged by challenges well beyond those of our own making. Not long after Alyssa's visit, the project was again subjected to yet another wild curveball; one that promptly dispelled the mundane and ignited our worst fears.

Welcome to the North Atlantic

May 19, 2000. Ocean Futures Society Press Release

Excerpt:

Ocean Futures Society learned that construction of a pier will occur in Vestmannaeyjar Harbor. The construction would involve blasting and pile driving at a distance less than half a mile from Keiko's bay enclosure. At this distance, the shock waves and low-frequency vibrations from the construction work could, in Ocean Futures Society's judgment, pose a risk of physical harm to Keiko.

Ocean Futures Society (OFS), through interactions with the U.S. and Icelandic authorities, was successful in delaying the blasting required in the harbor until at least May 25th. However, the explosives were already packed into bored-out rock in multiple locations, each several meters below sea level. The construction company's general manager pleaded with the harbormaster to allow the work to continue. Otherwise, the costly explosives would deteriorate from saltwater intrusion. The overdue need for harbor improvements, required by the fishing fleet within Heimaey and prior to the oncoming season, added urgency. Tensions mounted swiftly, pitting the economic lifeblood of the island against the visiting Keiko Release Project.

Charles, as the main liaison between the bodily protection of Keiko and grave threats posed by the harbor construction, found himself in a precarious position. Navigating the trio of conflicting objectives would not be easy. We couldn't knowingly expose Keiko to the deafening underwater blasts. His presence in the bay was only a few hundred meters from the blast site. On the other hand, delaying much-needed harbor expansion threatened to alienate the entire release project from the Heimaey community. This aspect did not bode well because we were so thoroughly dependent upon the small fishing village.

The most obvious solution, one we were ready for, was to remove Keiko from the bay during the blasting. Unfortunately, the official release permit required by Icelandic Fisheries (and U.S. authorities) was still in process. This was no routine permit, and the high-profile nature of the project on both continents placed an enormous regulatory burden on drafting and approving such a permit. Even if we were satisfied with the contents of our submission, evaluation and approval of such an intricate and unorthodox permit was unlikely to be fluid. Certainly it could not be completed in time for the imposing blasts set to take place a stone's throw from Keiko.

Anxieties quickly ran amuck within the ranks of the release team. When they first heard about the impending blasting, a few within the organization contemplated keeping the whale right where he was, questioning how truly threatening the explosions would be to Keiko. Others talked about "sneaking" our resident whale out in defiance of regulators if fast-track approval was not given.

Another solution proposed the use of an underwater air hose filled with pin-sized holes discharging high-pressure air into a "bubble net." In theory, the bubble net would create a wall of jumbled air between Keiko and the explosions. This prospect, though initially alluring, failed to address the shock waves that would be conveyed through the bedrock making up the entirety of Klettsvik Bay. There were even suggestions to construct a floating rig that Keiko could be trained to slide up on, temporarily out of the water and thus out of danger. A safe position to be sure, but who could guarantee that he would perform the behavior at the right time or stay in the position long enough to outlast the explosions?

None of the proposed solutions accounted for the unknown quantity of aftershock reverberations, and still others relied much too heavily on chance. To a person, not one of us supported any of the office-based elucidations hurled in our direction. Though they were well-meaning attempts at a solution, we knew without the need of forensic sound testing or scientific analysis that the blasts would be devastating within the parabolic-shaped rock echo chamber of Keiko's bay enclosure. We knew any outcome that kept Keiko in close proximity to the blast site was a nonstarter. The only alternative worthy of any discussion was taking Keiko as far away as we could from the dynamite and that left only one possibility, the open ocean of the North Atlantic.

As it goes with imposed threats and impending deadlines, the team pulled together against a common foe. Specifically, the task placed Jen and me in ceaseless collaboration to finalize the permit application. As luck would have it, I was on a preplanned yet brief escape in Ireland with Alyssa. To her dismay, I spent a great deal of our travels around the Emerald Isle on the computer reading daily iterations of the permit draft. Jen owned the research objectives, I the behavioral rehabilitation plan and release criteria. We conspired together on the historical aspects of the project spanning Keiko's time in Mexico to present. Robin and Jeff managed to piece together the medical history and flushed out the finer points of the all-important intervention portion of the permit, required by the authorities in the event that Keiko did not sustain the criteria which defined his successful release.

Days and nights melted together, one unrecognizable from the next. Overlying the constancy of permit drafts circulating this way and that, Charles worked with fevered urgency alongside Gummi and Hallur seeking varied strategies between authorities in both the United States and Iceland. In the eleventh hour, the OFS main office went so far as to publish a Web-based campaign. Their intent: to spread public awareness and thus gain government support in protecting Keiko. Here they posted near daily updates of the drama unfolding in Iceland. All the while the harbormaster continued to uphold his hardened assurance that the blast would go off on schedule, with or without our readiness.

The Piercing

Returning to Iceland amidst the escalating doomsday scenario, I was accompanied on the final stretch of my voyage by the notorious Dr. Lanny Cornell. Our paths converged outside of Reykjavik at the ferry terminal, the only means by which to reach Heimaey on that particularly foul day.

Throughout my entire career, which had begun on the heels of Lanny's unceremonious departure from SeaWorld, I had heard abundant stories about the infamous veterinarian. During those sixteen years in and around the small marine mammal community, I had yet to meet anyone who espoused a fondness for Dr. Cornell.

Though we had exchanged numerous e-mails throughout the course of the project (a healthy number of them contentious in nature), I had not met Lanny face-to-face since joining the release team. He rarely visited the project site, preferring to have clinical samples carried to and fro by the staff rotating in and out of the states—a practice that was not unlike smuggling contraband. More often than not we didn't have the proper documentation for biological materials transport. This was but one of many old-school attributes that characterized Lanny's detached husbandry leadership of the release campaign. Regardless, I was not a person with ready interest in conflict. Our first meeting in person was drawn out by the elongated wait typical of the ferry route.

He was sitting in one of the connected rows of seats inside the small terminal. I recognized him immediately. He wouldn't remember, but I had met him just weeks before his forced resignation from SeaWorld almost fifteen years prior. Without a doubt he had aged, though his square jaw and military-style short hair were unmistakable. That he was adorned in very American attire among the mix of European foot-traffic about the waiting area helped.

"Lanny?" I extended my hand. "Mark Simmons."

"Lanny Cornell," he replied, with little articulation, reminiscent of the stereotyped drill sergeant.

"How was your trip?"

"Long, but smooth."

He stood, folding the paper he had been reading. "Maybe we should get something to eat. We've got a long wait," he suggested.

Lanny was tall, at least matching my own height and maybe a fraction more. Despite his advancing age, Lanny was still an imposing figure. In his thick green winter jacket, his fully gray hair and robust frame added to a dominant posture.

In the terminal café, we each picked from the buffet-style assortment of unfamiliar foods then sat opposite each other at a small circular table. The chairs were more like comfortable reading chairs. Bound in imitation leather, they were low, making me feel like a child at the dinner table. My height was all in my legs. I felt absurdly small sitting in the rounded chair.

"So how's Keiko?" Lanny started.

"Doing pretty well. Better than I expected, to be honest." I didn't want to go into any detail knowing that I embodied the one aspect of animal sciences that Lanny could scarcely tolerate—that of animal training. But what I had heard of Lanny from the folklore within the marine mammal community was not yet evident. Although he maintained a businesslike seriousness, no evidence of the disreputable character emerged. I made every effort to avoid the disharmonious undercurrents that were most assuredly lurking just beneath the civil exterior.

Dodgy small talk wore on for the better part of our wait.

"I can't stand the ferry," I said, avoiding fuller topics. "After the flight I just wanna get to the island and be done with it. This waiting and the four-hour ferry ride drive me nuts."

Finally, as if he had been pining for the right moment, Lanny figuratively donned his ill-gotten crown, with pride and jest.

"Well," he said, sipping his coffee before continuing, "I decided to keep you company. Normally when the planes's not running I just walk to Heimaey." It was a creative way to align himself with

Christ, implying that he could walk on water. He left the comment lingering in the air with a straight-lipped smile.

"Ha," I croaked, looking at my shoes. "Well, I guess I appreciate the consideration." It was the least I could do. I had no witty comeback to offer.

By the time we boarded the ferry I had tired of the effort required in conversing without really saying anything in the process. I excused myself feigning the fatigue of travel and purchased a private cabin where I could sleep off the pitching four-hour ride to the island. Lanny did the same, and we didn't see each other again until dinner that night at Lanterna where we were joined by Charles and Robin. We were to discuss the project as a whole and the next steps required in getting Keiko safely away from the impending harbor construction. Lanny's presence was required to surgically attach the tracking tag to Keiko's dorsal fin.

Our evening discussion passed uneventfully. It didn't hurt that we each tempered ourselves with the customary red wine that accompanied a meal of lamb. Topics of weight had already been discussed over e-mail exchanges with Charles the mediator and go-between. The conversation over dinner was more a rehearsal of *tone* in the saying than any material change in content. Nonetheless Charles was noticeably more guarded, a manner I was not accustomed to seeing from him. It was as if he expected any moment there would be a clash across the dinner table, the instigator yet to be revealed. However, the dinner concluded without incident and we retired early for the night. Less a few minor jibes in small talk we were able to preserve the tenuous peace.

The next morning a respectable entourage amassed on the bay pen. We were going to attach the satellite tag, or at least a model of it. Brad Hanson had been in Heimaey just weeks before. Brad was an agent of U.S. National Marine Fisheries Service and had a lot of experience with tagging marine mammals. He was spearheading the development of a prototype tag for Keiko that we hoped would last much longer than the traditional VHF radio tags. This one was a satellite tracking tag, able to upload data when the long-arching antenna trailing behind broke the surface with Keiko's every breath. A VHF tag was a much simpler beast, but required a human to be in close proximity with a direct line of sight in order to pick up the intermittent signal. This would never do in the vast expanses of the islands and distances involved so far north.

Tinkering with the prototype, Brad and Jeff had produced several evolutions of the sat-tag, constantly trying to improve the design to its most hydrodynamic state. This wasn't an easy prospect: The "guts" of the tag consisted of the core satellite tracking technology, which was not small by any measure. The molded housing had to envelop the sizable mass of electronics while at the same time reducing the tag's drag. Jeff and Brad worked well together in crafting a workable model.

The two could not have been less alike. Brad, smaller of stature and cerebral, had a more academic poise compared to that of Jeff's rugged spirit. A focused seriousness permeated the immediate vicinity as he tinkered around Keiko or within the scattered mess of the Bassar storage building, which served as a makeshift lab. In the few and short interactions I shared with Brad, it seemed apparent that he did not take well to my particular brand of humor. He was often unresponsive to my descriptions of the tag as "the Volkswagen" or asking why we didn't just put a bright orange flag on Keiko, "Ya know, like the ones on a tricycle?" Lacking much in common, I often chided him with a dry humor in place of intellectual exchange. Brad didn't care much for the affronts posed at the expense of his baby. I couldn't blame him. At the end of it all he had worked tirelessly, determined to perfect the sat-tag housing.

The concern was not about the device slowing Keiko; it was all about reducing the migration of the tag. We knew it would eventually work its way off of Keiko no matter how well-designed, but the shape and location of the tag on his dorsal fin had everything to do with how long it would last. When swimming, water relentlessly pushing the tag backward meant that over time and distances, the tag and the pins would ever so gradually cut a path out the backside of his dorsal fin, leaving a swath of tissue damage in their wake.

The tag itself would sandwich the lower backside of Keiko's dorsal fin. The two sides of the tag would be joined by surgical-grade titanium pins approximately a centimeter in diameter. Given the size of the tracking device there would need to be three pins and therefore three holes drilled through Keiko's dorsal. Even at the trailing edge, so massive in size was his fin that the more anterior holes would pass through almost three inches of tissue. To start, we planned to remove the tag whenever Keiko was inside the bay thus preserving the integrity of the tissue at the place of attachment. In time, if Keiko stayed at sea, we hoped the design and stout connection would hold in place for more than a year, far longer than other similar tags of the time had achieved.

Tracking Keiko's whereabouts after a potential release was of paramount importance in evaluating whether he was thriving or not thriving. Emphasizing the importance, the ability to adequately track Keiko was also a clear requirement of the release permit. Without the tag in place, we would have no permit and no chance of getting Keiko away from the blasting. Just the same, the prospect of drilling through his massive dorsal fin with a handyman drill, and attaching a device the size of a small laptop was a procedure ill-suited for the faint of heart.

Weeks prior, we worked to familiarize Keiko with the unusual sensations of a battery-powered Dewalt drill, the very instrument that would be utilized to bore through the cartilage-like tissue in his dorsal fin. In these approximations, we held the body of the drill against his fin at the tag location allowing Keiko to experience the minute vibrations transmitted by operating the drill. We steadily increased the amount of time he was exposed to the odd sensation, reinforcing him for maintaining a calm disposition. As expected, the rehearsals were nothing for Keiko. By this time he had become apathetic toward the variety of strange contraptions his human counterparts often placed on his body. However, his relaxed acceptance of the drill practice did nothing to bolster our confidence. The actual drilling through his dorsal would be a far cry different than any simulation we could invent. Knowing this, we expected a struggle. In preparation for that struggle, we intended to forcibly restrain the 10,000-pound animal within the confining space of the small medical pen.

How does one restrain a nearly five-ton killer whale? Truthfully, it cannot be done, especially if the subject of the restraint is wholly unwilling. Rather, the idea is to provide the illusion of restraint, the "feeling" of being so thoroughly ensnared that escape is seemingly unattainable. Create helplessness. To do this, we would separate Keiko to the medical pool, eliminating any outward path of escape. We would then slowly position a killer whale-sized net lining the medical pool by snaking the net down one side along the bottom and up the other.

Advancing the plan, and while holding Keiko to one side of the medical pool, we would carefully draw up the net until finally it cocooned his massive body, supporting him from underneath and snugly wrapping his sides, as if a giant killer whale taco. The sensation of the net around his body, namely his head, flukes and pectoral fins (his steerage and drive shaft), should metamorphose the whale to a condition of apathy. In this position, Keiko would relax, the precarious and involuntary restraint further insured by the training staff taking every step available to encourage and prolong a calm disposition. In order to carry out our plan, we had all the muscle we could recruit present on the pen that morning, including the indefatigable Smari Harðarson. Unaware and apart from human machinations, Keiko had other ideas.

Separating Keiko to the medical pool and lining the pool with our trapping net went without a hitch. We immediately put Smari and Greg in the water in full dive gear. They would keep the bottom of the net from billowing which otherwise gave Keiko an easy out. The rest of us topside were charged with seining the net inch by inch until we had it, and Keiko, fully retracted. So it went. Keiko was held in a lineup position, parallel to the west side of the med pool along the surface. Robin, Jeff, Brian, Blair, Tom and I began crawling the net inward, heaving the sizable mesh and draping its excess onto the deck at our feet. As his med pool halved in size, Keiko, long familiar with nets, began a slow familiar dance with his old friend. He was not alarmed. He did not panic. He merely took a preparative breath, then submerging, slinked straight to the bottom. In doing so he was supremely calm, almost nonchalant.

On deck, those of us manhandling the net could feel we were on the losing end of a tug-of-war taking place somewhere in the turbid depths of the medical pool. Imagining what might be transpiring on the bottom, we were reluctant to put our backs into the struggle. After all, we had two divers in the water. However, releasing the net would only make a potential problem turn into a real emergency. Any slack afforded could just as easily give cause to a mess of whale and divers entangled in a tightly wound and unforgiving ball of tensioned havoc. In those scant few moments, the devastation that would result if Keiko decided to thrash and spin within the trapping mesh played like a nightmare unfolding in our minds. It seemed we were walking on the very edges of our worst fears. At any moment the pall of uncertainty would emerge into a catastrophic upwelling of white water, divers, net and whale.

Somehow fortune favored us that morning. More accurately, Keiko's experience in avoiding nets paid unexpected dividends for our group of wary stakeholders. By now, the opposite topline of the encircling net had divided the medical pool in half; the worthless excesses lumped at our feet on the west side of the deck. Shortly after the mysterious goings-on at the bottom of the medical pool, Keiko emerged at the surface on the east side, clearly free of our ingenious compass.

"What the heck?" I blurted.

"Can anyone see Smari?" Robin called out. Greg had come to the surface just moments before, obviously the wiser.

"Greg, where's Smari? Jeff repeated Robin's concern.

"I can't see anything down there. It's so stirred up," Greg replied.

As his words still hung in the air, Smari popped his head from the surface on the south end of the medical pool, like Keiko, outside of the net. He looked both frustrated and astonished.

"He got out," Smari said, as if he were the first to know.

"How did he . . .?" I was stumped. The entrapping net was overly large, easily covering the extents of the medical pool. We had carefully maneuvered the net completely enclosing the entire pool, no corner left unanswered. I couldn't imagine how Keiko got through.

"Smari, you and Greg go ahead and get out," Robin urgently instructed.

We stood poolside, for the moment leaving the impotent catch net loosely waving in the light chop.

"Let's just get the net out before someone gets hurt. This isn't going to work. He's too net-smart," Jeff said.

At that, we pulled the useless net clear of the medical pool, piling it in a balled-up line along the west deck. Keiko floated on the far side making no attempt to solicit our attention. It felt almost as if he had gone to his corner, awaiting round two. Smari came over the bridge and approached the brain trust leaning forward as he walked. He was still wearing his full dive gear.

"Holy crap, man!" he said, his Icelandic pronunciation making "man" sound overly innocent. Gesturing out a very small circle with his hand, he continued. "He found a tiny gap at the edge and pushed the net and lifted it up. I tried to keep him back. I was pushing on his nose and trying to keep the net down, but, man, he just came through."

Smari was competitive. He didn't like being the guy that let the whale get away. The outlines of his dive mask on his face and reddened eyes lent to his frustrated appearance. "I can't believe he could get his body through that tiny hole," he said, as much sheepish as exasperated.

"Ohmygosh, dude, you're nuts! You can't beat a whale." I found the prospect humorous, but only because we were safely clear of the potential mess. Smari's one-on-one battle with Keiko only evidenced his inexperience. We were very lucky. Once gone that far, it was fruitless to attempt to forestall Keiko's escape, and the effort to do so was gracefully excused by the all-too forgiving whale. Any other animal, and we might be cutting a dead body from the net right then, human, whale or both.

"Why don't we just ask him to hold voluntarily?" Jeff proposed. "He's been pretty good with letting us work on him in the past."

As always, with a tone of optimism, Jeff offered the path of least resistance. Admittedly, Robin and I had been proponents of netting Keiko for the procedure, but then again, Keiko was no ordinary whale. His passive acceptance of what would piss off an otherwise normal whale lent to our initial misdirection.

"It's about the only option we have. There's no way we're going to get him in the net after that. He beat us, and he knows it," I replied. "He pulled that move like a pro."

So it was, we would simply ask Keiko to line up at the surface, hold him in position and allow the doc to do what he needed. A short while later, having allowed the atmosphere to settle, I stepped up to the medical pool and asked Keiko over. All seemed well enough. We lined him up alongside the HDPE pipe of the west medical pool. Here, Tom and Tracy took position at Keiko's head. They would keep his focus and periodically reinforce him for staying in position. Brian and I moved down to his dorsal. The massive girth of his body required that we pull him inward, anchoring his midsection as close to the pool edge as possible. This enabled Lanny to reach Keiko's dorsal fin from poolside. Hunched over, our knees pressing into the Chemgrate decking, Brian took the leading edge and I the trailing edge of Keiko's dorsal. Lanny squeezed between us, sitting comfortably on the deck with his boots resting on the HDPE pipe.

First, he injected Keiko's fin with a numbing agent, likely carbocaine. This took some time. Injecting solution into the very dense tough tissue required no small amount of hand strength. The grip required to keep Keiko's dorsal fin where we needed it was already cramping both my hands. Lanny applied the local anesthetic on the two sides in a scattered pattern surrounding the three areas to be drilled. Normally, five or more minutes is needed to allow the numbing agent to fully take effect. By the time Lanny finished the last of his shots, already ten minutes had passed since the first. He asked for the template and the drill and went straight to work.

Earlier Jeff had gotten in the water, propping himself on the outside of Keiko, in line with Lanny, the dorsal fin between them. He supported himself on Keiko's back, his left arm hooked onto the front edge of the whale's dorsal fin. Submerged to his chest, the air in his splash suit was forced to the top like a half-empty and rolled-up tube of toothpaste. He looked uncomfortable and not a little ridiculous. Jeff would help keep the template in place and guide Lanny on navigating the drill's alignment straight through the dorsal. It was important that the hole match on both sides.

Robin stood just behind Lanny preparing the drill, which was equipped with an expensive diamond-edge bit, though nothing different than what can be purchased at The Home Depot. After sterilizing the bit Robin handed the drill to Lanny. The infamous doctor did not hesitate. His approach was shocking . . . I had taken more precaution when building a deck on my house the previous summer. The razor-sharp bit dove right through the outer skin layer with no noticeable effect on Keiko, who sat almost motionless. As Lanny got into the heavier cartilage, the progress slowed, but there was no grinding sound, nothing grotesque about it. It appeared and sounded like drilling through wet balsa wood. Within seconds Lanny had completely perforated two and a half inches of Keiko's dorsal. The first hole was the more forward of the three attachment points. But Lanny had missed the mark. The path of the drilled hole did not exit in line with the intended target on the other side. He was off by more than a centimeter. By then the tiny vessels within the connective tissue had begun to bleed. Crimson red flowed steadily from the gaping hole, down Keiko's dorsal and into the water. Diluted and spreading within the water column, the watery red cloud appeared as if it were a cheap special effect in a third-rate horror film.

Lanny moved the template higher and remarked the same pin location. Without a word he began drilling again. Though we all shared the same inward trepidations, no one questioned Lanny's quickness to pierce yet another hole in Keiko's dorsal fin. For the most part, we all expected that the veterinarian, of all people, would naturally espouse more exacting restraint than the average Joe. After completing the second hole, again missing the alignment, Lanny began boring out the same in an attempt to fix the erroneous path. At this Keiko's patience began to dissipate. Whether he was feeling pain or just curious, it was impossible to know, but it was abundantly clear he wasn't going to hold his position much longer. I was relieved. Lanny's come-what-may approach to the task cast an air of astonishment and disgust among those of us watching the crass process in which he callously drilled and drilled again.

Of his own volition, Keiko dropped his body submerging his dorsal and turned, sitting more upright in the water and directly in front of Lanny, who remained in the same sitting position at poolside. I stood and looked at Tom and Tracy, who had nothing more to offer than an impotent shrug. They had done all they could do to keep Keiko lined up, but he had reached the limit of his patience regardless. Still uncertain and more than a little perturbed at Lanny's handling of the "surgery," none of us was quick to offer direction. Amidst our pensive hesitations, Lanny reached back into one of Keiko's buckets and tossed a single herring into the whale's mouth.

"We need to break," I was instantly pissed. "We need to break *Now!*"

"What do you want to do?" Lanny asked.

"I don't care, but you need to step away from the pool right now. We need to clear the area." I stepped back, trying to draw the entourage surrounding Keiko with me.

The entirety of the last twenty minutes coursed through me at once. I hadn't made any attempt to stop it. I was angry with myself. I was vexed by Lanny's cavalier approach to what should have been a precise and calculated procedure. To make matters worse, he had just reinforced Keiko after Keiko had prematurely ended the session of his own accord, an incorrect behavior to be sure. That was not the only offense. Lanny was no one that needed to be associated with primary reinforcement and further, gave no heed to the behavioral regimens and principles we had worked on for over a year. The doctor acted innocent at the gesture, but he knew exactly what he was doing. It was a passive-aggressive affront toward everything I represented.

"He's sitting calmly," Lanny insisted, as if to justify the slight.

I ignored it, looking instead at Robin, who knew I was about to explode.

"Let's get everybody in the research shack, away from the pool and out of sight from Keiko," Robin intervened.

As everyone moved inside, Robin and I stepped around the south end of the research shack. He knew I was aggravated and swiftly created a diversion by requesting my input on the next steps. "What do you want to do?"

"I can't believe that shit." I wasn't ready to answer yet, I just wanted to vent. "We can't do that again, and if he touches another bucket I'm going to put it over his head."

"He won't and I won't let him," Robin assured me, his tone was casual.

"What about drilling willy-nilly . . . like he has no idea what he's doing?"

"Jeff and I will talk about it. I'll suggest that Jeff finish."

"Well, we've already made a mockery of this entire situation. We need to step back up with minimal people, who know what the hell they're doing and keep this short. Even if that means doing it five more times," I said.

"Let's get inside and talk about it with Charles, Jeff and Lanny," said Robin.

Back inside, the mass of bodies shifted uncomfortably about the confines of the research shack, which was not intended for a group of this size. All of a sudden it was hot. Coming from outside and in full gear to a heated and overpopulated trailer made me directly aware of the bulky splash suit. I felt like a kid in six layers of snow gear that hadn't even left the house yet. My lingering angst contributed to the discomfort in no small part. In the close quarters of the shack, Lanny started in, poking the bear.

"Every project like this needs at least one genius and one asshole." Lanny said smugly.

He knew I would be easy to unravel. I welcomed the liberty. "Good, then you can be the asshole." After the briefest pause, I finished the assignments. "I'll be the genius." I turned facing Lanny as I made the comment, pulling the tight rubber seal of the splash suit away from my neck in the process.

Like a seasoned referee, Jeff interrupted the exchange offering an alternative. "I think I can get it. I think I can get a better angle from the water." It seemed obvious that he and Robin had somehow already conversed on the topic. That, or Jeff was thinking along the same lines as Robin.

He continued, "It's hard to get the angle right from where Lanny is, but I can get a much better view from where I'm at. I think we can make it work. I just might need Brian or someone to hold me up when I'm working the drill."

There it was. Jeff skillfully and diplomatically took over the surgery. He had given Lanny an easy out, inasmuch taking the gun right out of his hand in one fell swoop. Agreement settled, I preached my bit on the approach and crowd control. Immediately after, those of us charged with control of the session started out of the shack. As we shuffled through the crowd and back out onto the bay pen, I mumbled under my breath to Brian, Tracy and Tom, "Make sure there are no buckets within Lanny's reach."

The morning wore on, and we completed the attachment in only two additional sessions, each of which far surpassed the first in efficiency, accuracy and all-around behavioral correctness. Not much else was shared between Lanny and me that day. For the most part, I remained quietly on the fringe. Though admittedly, I kept a careful eye out and almost hoped he would cross a line.

Perhaps my shortened fuse with Lanny was largely vested in his tainted reputation. Just the same, there was no lack of present offenses with which to find fault either. I knew what dedication to the animal looked and felt like. I witnessed nothing redeeming in Lanny's skillsets that granted him the grace and favor of his self-appointed title as Patriarchal Asshole.

The following morning Lanny left the island. As quickly as he came, he went. The whirlwind nature and term of his visit did nothing to assuage already deeply ingrained perceptions of him among the staff. Mercifully, he had been nothing more than a passing inconvenience, rarely on-site and seldom affecting any true influence. Still, more than a few of us breathed a welcome sigh of relief at his departure.

Glimpse

Finally, dangerously close to the ill-fated blast date, Charles reached a compromise with Icelandic Fisheries. It was a brilliant move on his part, offering the reluctant officials an olive branch. Rather than a blanket approval, a stipulation was introduced such that Keiko could go to the open ocean but with limitations. He could not be introduced to other killer whales, at least not until other components of his release criteria could be objectively measured and proven.

As the summer season brought with it extended days and often calmer waters at sea, so too did it bring the migration and nearby presence of many whales. The approval meant we now had the means to escape dangers in the harbor, but it also meant excursions to sea could develop into a dodgy game of cat and mouse.

It was agreed, we would evade the blasting in the harbor by taking Keiko to sea and in doing so, move the project purposefully into the next phase of reintroduction. It seemed we had only yesterday been sequestered in the bay pen struggling to overcome limitless obstacles, and yet here we were: Keiko a different animal, the staff getting comfortable with the reality of "letting go" and the imposing threat of nearby blasting. This was the unexpected plot that would see us off to the next adventure in Keiko's journey to freedom.

Just days before the blast date, we had our tempered approval to take Keiko out of the bay fast in hand. He would be free from any manageable form of confinement for the first time in twenty years. Though the goal of the project was indeed release to the wild, the prospect of Keiko outside the bay initially caused much uneasiness among the staff. After all, none of us expected him to just swim off, but we also didn't know what to expect. For starters, we didn't know with any certainty that he would stay with the walk-boat. There was also concern that he might become disoriented in the new surroundings and go directly into the harbor, closer still to the dangers we sought to avoid. Reflecting on the tribulations of his first access out of the bay pen itself, a very real possibility existed that he would not even follow his familiar escort, refusing to leave the bay altogether.

No time was wasted. Only two days remained before the blasting would take place. Upon confirmation of the staged permit, we immediately set up a practice session. Before attempting to take Keiko far out to sea, we would simply take him outside the bay by sending him from the bay pen platform through the unfamiliar barrier net gate and to the *Draupnir* waiting in the shipping channel; a onetime dress rehearsal. The test was important to calm our own anxieties more so than anything else.

On the day of the preliminary test, suppressing summer light and resulting warmth helped to calm the otherwise fitful elements. We had enjoyed a run of agreeable weather. It almost felt as if the Nordic gods were sympathetic to our circumstance and welcomed the next baby steps in Keiko's heightened adventure.

The *Draupnir* idled a hundred or so meters outside the barrier net gateway, starboard platform facing the direction of the gate. Onboard *Heppin*, Greg and Smari positioned on the seaward side of the channel to run interference should a third-party vessel unknowingly motor into the peculiar scene. Likewise, they were ready to move about and intercept boats coming from the harbor side. *Sili* was tied up to the barrier net itself, prepared to open the underwater guillotine-style gate. Tom and Brian were on the bay pen. They prepared to send Keiko to the awaiting *Draupnir*.

On the receiving side, *Draupnir* carried a full crew. We had to be prepared for any possible outcome. Jen was stationed atop the pilothouse, recording gear in hand and wearing a red helmet mounted with camera to capture data for Keiko's first miniature exposure to the open sea beyond. Michael captained the walk-boat and was accompanied by Charles inside the cabin. Robin positioned himself directly behind my usual stance on the already extended platform. He conducted the sequence of events from his handheld radio.

"I think we're ready," he said almost to himself, as he scanned across all boats and positions. Like many of us during our rotations to the remote island, he had let his beard grow out. The fully gray scruff lent itself to a salty seafarer guise.

"I'm ready here whenever they are," I offered in response. Tracy was beside Robin, there to assist me with whatever I might need from my perch on the platform. One hand was always holding fast to the platform. Every movement of the boat was exaggerated on the artificial appendage.

"*Draupnir*—Bay Pen," Robin called over his radio.

"Bay Pen. Go ahead, *Draupnir*." It was Tom who responded, which meant that Brian would be working Keiko.

Robin continued, "Call him over when you're ready. You guys will have to call the gate."

Robin was reminding them that the gate movement was always dictated by whoever was working with Keiko.

"Robin, Blair and Dane know to leave the gate open, right?" I asked of no one in particular. Dane Richards, Lanny's nephew was part of the operation. We'd been over and over the sequence back at the dock, but one could never be too careful when it came to undiscovered country. Even the slightest mistake could cause an avalanche of unwanted results.

"They know," Robin replied.

"Bay Pen—*Sili*," Tom called over the radio. "Open the gate."

Keiko was sitting up in front of Brian, who was kneeling on the matching platform extended from the south ring of the pen.

"Copy that, Bay Pen, gate opening," Blair responded.

We watched as Blair and Dane unlashed the lines holding the net gate in its closed position and slowly lowered the invisible gate panel, feeding out the line hand over hand.

Moments later the "gate's open" call came from the *Sili*.

"*Draupnir*—Bay Pen. Send 'em," Robin instructed.

Even before Tom responded with the confirmation, we could see that Brian was standing and pointing Keiko directly toward the barrier net gate, one arm extended straight out from his chest. I waited just a moment, to be sure Keiko's head and thus his ears were submerged.

"Tracy, recall."

"Recall," Tracy repeated, confirming that she had hit the tone.

I immediately followed her tone with a slap of the target, hitting the water like I was trying to knock a baseball out of the park. I wanted to be sure Keiko could discern our bizarre position outside the bay. I waited until I thought Keiko would be close to the gate and slapped with the target a second time, just as forceful as the first. Keiko was nowhere to be seen, although this was to be expected. After all he had to submerge to pass through the gate which was located a few feet below the surface.

Robin radioed, "Call it if you see him."

No response came and still no sign of Keiko.

"Hit the recall again," I asked Tracy, preparing to follow her acknowledgment with a third slap of the target. As I rounded up to strike the surface, I had to pull back at the last second nearly hitting Keiko in the head, who had come from nowhere and popped up in front of me. He was completely at ease. As far as he was concerned, it was nothing more than another walk rehearsal. Exaggerated by our anxiety, it had seemed an interminable wait before Keiko finally reached the *Draupnir*, though he actually covered the expanse in less than a minute. Outside the sanctity of the bay, we were all one interconnected ball of nerves. Keiko was happily oblivious. Change had become his norm. To him it was likely just another odd game his human friends dreamt up.

I offered him two or three herring by placing them below the surface of the water. In over two months, we had no longer given Keiko food directly in his mouth or for sitting up above the surface with his mouth open. A small but logical step, all food of any kind was given beneath the surface, even when it came from human hands.

"I'm ready to point him back," I said as I stood and looked down at Keiko. We had purposely set the rehearsal up for mid-tide. As expected and hoped, the water was near flat calm requiring little effort from Keiko to stay at position off the *Draupnir's* walk platform.

"*Draupnir*—Bay Pen. We're pointing him back." At Robin's call, I pointed my left arm in the direction of the bay pen. Keiko lifted his head exposing his white underside and slowly rolled back and away from the platform.

"Slap!" Robin called on the radio.

Repeating the exact same sequence, Tom and Brian hit the recall tone at the bay pen and followed it with a target slap. After roughly the same passing of time, Keiko showed up in front of Brian at the bay pen.

"Awesome . . . that was excellent!" I couldn't contain my momentary relief.

"Good boy!" Tracy shared in the moment.

"*Draupnir*—Bay Pen," Robin was moving on already. "Hold off on your food. We're going to do one more, this time we'll move a little. Save the majority for the last sep to the pen."

It was Tom's voice again over the radio, "Copy that, *Draupnir*."

Turning to Michael but touching my shoulder as if to include me in the audience, Robin said, "Michael, once we get him, we'll poke our head outside the mouth of the channel, then come back to this same position and point him back inside the bay."

"Got it, boss," Michael replied, referring to Robin in his customary way.

"Robin, I think we should end from the *Draupnir* inside the bay," I said. I didn't want to keep asking him to go away from his favorite toy. I felt the best reinforcement for returning to the bay was to get his walk-boat back, even if just a touch-and-go.

"Agreed." Then turning back to Michael, who was splitting his nervous attentions between the shipping channel and Robin, he said, "After we point him back into the bay, we'll close the gate and take the *Draupnir* into the bay."

"Whatever you say. Just let me know when."

By now, Michael knew well not to move the *Draupnir* without clear direction from the training staff. At times our demands on the captains of the formation were at best unusual, at worst completely mystifying. They each knew better than to try anticipating what would come next.

Keiko was pointed back out to his walk-boat. A mirror of the first "A-to-B" exchange, he arrived at the *Draupnir*, following our lead with seemingly blind faith. This time Jen spotted him on his approach from her position in the makeshift crow's nest above the pilothouse.

Advancing the plan, we motored ahead toward the mouth of the channel bordering the wide open North Atlantic. Closer to the mouth, the surface swells of the bay amplified into actual waves, nearly three or four feet at their caps. Sitting on the platform was no longer possible. I had to stand, holding tight to the forward guide rope secured high on the *Draupnir's* pilothouse. Keeping my knees slightly bent, I flexed my legs in rhythm with the rise and fall of the platform over each wave and contrasting pitch of the boat.

Keiko followed in his typical position and except for the embellished porpoising required to clear the waves for each breath, he gave no indication of anything out of the ordinary. We followed a wide arching circle tempting the open expanse of the northern seas, then were quickly back in calmer water at our original position off the bay gate. Less than ten minutes had passed.

"Piece of cake," Tracy piped triumphantly.

"Watch it, sister, we've still got to get him back to the bay."

"He'll be fine," she countered, energized from the brief exercise.

"*Draupnir—Sili,*" Robin radioed. "We're going to point him back to the bay pen. This time we'll close the gate. Bay pen will call the gate." He knew that Brian and Tom were listening to the transmission.

Blair acknowledged, "*Sili—Draupnir,* copy."

Keiko arrived at the bay pen platform without incident. Not realizing it, I took a larger than normal breath. Tracy heard my involuntary relief.

"What'd I tell you? See, you need me here to keep you from going crazy."

"You're the one that makes me crazy."

Our verbal joust was evidence of a relaxing posture onboard the *Draupnir*. Amidst our exchange, we heard the familiar call of the gate and watched as Blair and Dane pulled the guillotine gate back into closed position, tying off the leads when they were done. Keiko didn't so much as cast a glance in the direction of the rattling gate.

"*Sili*—Bay Pen. Gate secured."

"Affirmative, *Sili*. Thanks," Tom replied from the pen.

"Michael, let's move the *Draupnir* inside," Robin voiced. Then from the radio, "*Draupnir*—Bay Pen, we're coming over the boat gate. Inside the bay we'll do a short walk once or twice around the pen and call it a day."

"Copy that, *Draupnir,*" crackled back over the radio.

We concluded the approximation with a routine walk within the bay. This time we delivered the majority of Keiko's primary reinforcement—food—back in the bay enclosure. For now, it was important that we insure his reliable return. In the very near future, this balance would deliberately be reversed.

The dry run had gone almost perfectly. From Charles down, we were full of ourselves at the success of the day's trial. At least for this one brief night we would not have a care in the world. That evening, the red wine flowed in abundance.

Mark A. Simmons

Eight Nautical Miles

As the saying goes, "Time flies when you're having fun." It's even faster when bookended by mounting pressures and looming deadlines. Our short night of celebration quickly faded to a fleeting memory. The following day we had much to prepare for the full-scale walk to sea. Blasting in the harbor was set to take place in less than thirty-six hours. The next walk was not a rehearsal. We would go much farther than the mouth of the channel and be at sea for an unknown length of time. It was one thing to circle around within the shipping lane and another entirely to circumnavigate the island. Literally and figuratively, the farther we got from the relatively predictable bay, the more variables we could encounter. Any concern over Keiko was upstaged by apprehension of third-party vessels, weather and currents. Added to these somewhat unmanageable variables was the overriding directive to not expose Keiko to other killer whales.

May marked the beginning of the seasonal presence of many black-and-whites in and around the chain of islands, and no one knew how far away from other whales was far enough. It would take all of our man power and constant communication to ensure that we didn't stumble unexpectedly into a pod of orca as we rounded Heimaey away from the blasting site.

May 25, 2000: Weather continued to hold in our favor. Shortly after first light, everyone met at the harbor. A few were clinging to their morning coffee mugs to warm their hands. Others were fast at work transporting pelican cases. And our tired little red truck was making busy trips to and from the hotel and the fish warehouse. The uncharacteristic early morning hustle and bustle of loading boats, delegating assignments and last-minute checks created an atmosphere uncommon on the project.

Stubbornly, the inspired vision of Keiko swimming off at his first chance of escape lingered. The unspoken insinuation was tangibly evident by the determined resolve to document every element of the day, and more so, by sentimental reflections and bold wagers shared in hushed sidebar conversations. Dispelling everything that had preceded this bold new day in the project, some scant few still believed more in the Hollywood adaptation than that of the storied life which lay before them.

It was 0830 hours: The *Sili* set out of the harbor first. She would transport Brian and Tom to the bay pen, then exit the bay and assume her position on the barrier gate as she had done just two days before. Next out was *Heppin*, piloted by Greg and crewed by Smari, our fearless Icelandic head of security. Any third-party interference at sea was his responsibility.

Onboard *Draupnir*, packed to the gills, were Robin, myself, Tracy, Charles, Jeff, Jen and Michael. Keiko's food for the day occupied almost the entire aft deck space, crowded around the engine compartment in several well-iced buckets. Outside of the amidships cabin, the aft deck constituted the driest portion of the boat. Emptied "pelican" cases left from camera and recording gear were nestled inside the pilothouse and on the foredeck. Navigating the length of the *Draupnir* was made possible only by walking on her sponson while gripping the pilothouse, lest we be thrown overboard in the pronounced pitching of the boat while at sea. No matter the weather, this was the North Atlantic and her vast depths translated into a highly articulate surface even in the best of conditions.

Except for Charles and Michael, everyone wore his customary splash suit or Mustang survival suit, the only attire that would keep a person alive long enough to be rescued from the frigid waters. The bright yellow or bright orange of the survival gear bespeckled the decks of each craft, granting a semiformal appearance to the nonuniform flotilla of vessels making up the walk-formation.

Siti, his father and Ingunn comprised our land-based lookouts, filling the waterborne gaps in communication. Their primary responsibility was to spot killer whale pods in the vicinity of Vestmannaeyjar and relay their location and heading to the *Draupnir*. They would also provide cell phone backup if by chance the walk formation was out of marine radio range from the island base.

Finally, perhaps the most anticlimactic yet vital position in the chain of responsibility was standing alongside the harbormaster and coordinating the exact moment of the blasting. This duty was entrusted to the wise and reliable diplomacy of Gummi. The plan did not stop short of distance alone. Taking no chance, we would ask Keiko to spyhop at the instant of the blast to doubly ensure that the event would not spook the Big Man while so far from our base sanctuary. Gummi had to communicate that moment with precision.

Identical in form and function, we began Keiko's first adventure to his home waters just as we had rehearsed two days prior, only this time we did not stop at the mouth of the channel. *Heppin* was our scout, ensuring the way forward was clear of shipping traffic and/or killer whales. She advanced out of the channel and immediately turned to port on a north-northeast heading, roughly a half nautical mile off our bow. The *Draupnir*, with Keiko in step, emerged from Klettsvik next, crossing the disorganized chop at the top of the channel. After securing the gateway and picking up the bay pen crew, *Sili* brought up the rear of the formation. The least seaworthy of all the boats, the *Sili* would be an exhausting ride for her passengers.

As we rounded the northern most point of Heimaey, the waves turned to a following sea that slightly outpaced the *Draupnir*, the distance between the peak and valley of each almost twelve feet. Without skipping a beat, Keiko moved out, away and clear of the *Draupnir's* lee, riding the sizable swells on each downward run.

From my position on the platform, I had the most remarkable view of Keiko's first foray in the wild blue yonder. Immediately a natural, he swam with the current on the rise of the swell, then placing his flukes in an upward position, sailed effortlessly down the other side. The vision was surreal. Within the vastness of the clear surface and deep blue depths beneath, Keiko was again dwarfed by his surroundings. Watching him on the waves, more proficient at each repetition, our charge looked as if a child on his first playground slides.

"That's unbelievable. He's totally riding the waves!" I exclaimed to anyone within earshot.

"Man, you gotta love that. He looks right at home," Michael added through the open porthole of the pilothouse. "Look at him."

Peering over his shoulder up at the crow's nest, Jeff asked, "Jen, are you getting this?"

"Yep, although it's going to be nauseating to watch," she replied as she tried to steady herself and thus the helmet-mounted camera. In her position atop the pilothouse, the fore-and-aft pitch of the *Draupnir* was greatly exaggerated.

In the background, Michael radioed to the support boats, describing the scene for their mutual benefit. The engine pitch of the walk-boat vacillated between the uphill effort followed by a downhill idle as she navigated atop each passing swell. Keeping her on a straight line was not easy, as the following seas continually attempted to turn her broadside to the waves.

Turning further to port, we steadily made our way to a more northwesterly heading, staying only roughly half of a kilometer from shore. At this, the swells breaking around the eastern extents of the island laid back down, the following current gradually faded to a minor chop in the northern lee of Heimaey. Keiko then had to work to stay abreast of the walk-boat.

Thus far, we were holding about four to five knots. As the walk turned from wave-riding enrichment to slogging exertion, Keiko slowed and repeatedly gravitated to the underside of his man-made escort. Here he could ride the *Draupnir's* slipstream created by the hull pushing through the water. This would not do. If he were to survive the open ocean, he would not have a crutch by which to traverse the grand distances required for survival.

"Robin, can you reach the target pole?" I asked, my free hand outstretched in anticipation. "I'm going to reposition him, keep him about ten feet from the side where he can't cheat off the boat."

"Let me know when you want me to toss fish," Robin relayed. He knew I would not be able to simultaneously work the target, hold on to the platform's topline, and toss food to Keiko.

"Okay, on the first bridge for sure, then I'll stretch it out. We've got a long ways to go."

At the prompting request of the target, Keiko obediently shifted position and touched the small buoy with his rostrum. The achievement was not made easy by my clumsy effort to hold the target steady between the chop and the awkward pitching of the platform. I bridged, and Robin immediately tossed a herring about six feet in front of Keiko. He completely ignored the herring that swiftly disappeared in our wake.

"Let's try again. I'll get the target completely clear before you throw the herring so he's not focusing on me."

On the second attempt Robin threw the herring with more force, causing a better disruption in the surface when it landed. This one Keiko grabbed.

We carried on this way for the better part of four or five nautical miles. At random intervals, I prompted Keiko's position further out from the *Draupnir*, anticipating when he might attempt to ride in her slipstream. Periodically, we'd provide the occasional herring or two, but mostly we only gave the familiar "thanks" of the whistle bridge. We needed to save the majority of Keiko's feast for his return to Klettsvik. We also learned that a slower pace was more conducive to keeping Keiko in the proper position off the starboard side of the boat. At lower speeds, there wasn't much of a slipstream by which he could take advantage.

Shortly after 1030 hours, over an hour at a steady pace of three to four knots, we reached the northwestern most point of the island, our plotted destination for the blast avoidance. The *Draupnir* took up a stationary position but did not assume the neutral stance of encouraging exploration. This time, we kept Keiko at the platform in preparation for the blast in Klettsvik. *Heppin* and *Sili* stood off approximately a quarter nautical mile in opposite directions. The crew of each vessel was intently watching the watery horizon surrounding the vicinity for other boats or killer whales. So far so good. The water's surface was calm enough to make sighting of other marine mammals ideal. Our land-based lookouts spotted a small pod of wild whales during our trek. Luckily, they had been on a steady northern heading moving well away from our location.

Charles was on the cell phone, presumably with Gummi, coordinating our readiness and awaiting the countdown from the harbor master. From the extended platform, I put Keiko through the paces, requesting a battery of voluntary husbandry behaviors. The exercise was partly a test of his attentions in the new environment and partly to keep him focused on me. Not knowing the exact timing of the blast, I needed to keep him close and ready to engage in the spyhop we had planned.

"One minute," Charles advised Robin, holding the cell phone to his ear.

At that, I held Keiko close by the platform, his rostrum on the flat palm of my hand, targeting him in a ready position with his head up and above the surface. This forced him to move his body into a vertical position.

"Thirty seconds," Charles continued.

"Count down from ten please," I requested. It would take me that long to get Keiko to respond to the target.

"Ten—nine—eight—seven—"

Extending the target out from the side of the boat, I tapped Keiko's rostrum lightly and moved the target no more than a foot and a half above his head. The request was high enough to clear his ears above the surface, but not so high that he couldn't hold the position for a sustained few seconds.

"Four—three—two—"

The timing was just right. Keiko responded in coordination with Charles' relayed count and rose from the water touching the outstretched target. The brilliant white of his underside lit up in the sunlight as he reached high enough to bring the top third of his giant black pecs out of the water. He held the position, undulating his body to remain in contact with the target, dutifully awaiting the expected whistle bridge for a job well done.

"Okay?" I asked, knowing I didn't have much longer before Keiko would end the behavior. It couldn't be easy to propel his hefty mass to this position and hold it for long.

"Stand by—yes, the blast went off," Charles finally responded. Only a few moments had passed. Upon Charles' confirmation I bridged Keiko and tossed a few of his favorite fish around the water.

"No reaction whatsoever," I relayed to no one in particular.

"*Sili*, anything?" Jeff asked over the radio. The crew of the *Sili*, stationed between us and the return to the harbor, had submerged an underwater microphone and listened intently for any evidence of the blasts.

"*Sili—Draupnir.* Nothing on the hydrophone," came Blair's reply.

"Well, that went about as smoothly as we could hope for," Charles stated as he leaned over the sponson beside Robin looking down at Keiko, who seemed happily oblivious. Dutifully, Michael again relayed the update to the two support boats orbiting the *Draupnir* at a distance.

By this time, nearly every person on the boat was packed to the starboard side in the optimal position to observe Keiko at the time of the explosions. The event couldn't have been more anticlimactic from our distant position northwest of the island. Relief at finally evading the doomsday blasting that had created such a panic now erupted into a cackle of small talk aboard the walk-boat. As the dark cloud of the blasting lifted, the realization poured over us that here we were, out to sea with Keiko for the first time. Taking in the scene, afforded us on a gorgeous day, no less, we each took time to bask in the glory of the moment.

Not wanting to remain exposed much longer, Robin instructed Michael to prepare for the return. Following a few exchanges of position and route between the walk-boat and supporting vessels, a phone call or two to the island base, and bustling about the deck, we started off on an east-northeast heading, back to Klettsvik. It was barely after 1100 hours. The warmth of the sun, now high on the horizon, seemed to smile on our little accomplishment.

On the journey back, the currents that parted their way around the eastern point of the island did much to slow our approach. Keiko labored alongside the *Draupnir*. At times he seemed almost reluctant, forcing Michael to slow the pace across the small chop paralleling the northern shores. Unlike walk rehearsals in the bay, Keiko was all business at sea. Though he followed my direction when it was offered, he rarely cast his normal sideways glance in my direction. Instead, he kept his head down, powering his flukes rhythmically alongside the *Draupnir*.

Robin and Michael conferred back and forth on a tactic, a means to avoid the more amplified current and swells breaking around the rocky points guarding the entrance to Klettsvik. The two decided to run far and wide, making our turn south well beyond the tip of the island.

I could hear the conversations behind me on the platform, although my attentions remained on Keiko and holding fast to the line securing the erratically pitching platform. I wondered if my legs would be sore after what amounted to several hours at this position outside the boat. The ride was not unlike the twisting impacts of skiing down a black diamond slope, enduring moguls along the way.

Robin, Jeff and Charles were discussing the prospect of getting a blood sample from Keiko. By taking the sample then, the markers indicating Keiko's blood oxygen levels and other indicators of physical fitness could be measured more accurately. If we waited until our return, our star athlete would already be in recovery from the exertion.

Reaching south, we finally came into the lee of the barrier island Bjarnarey, where the wind and surface currents laid down. Michael brought the *Draupnir* to a full stop but kept the motors running and hovered her, to the best of his ability, in a stationary position.

Behind me, Robin and Tracy prepared the butterfly needle and syringe as I asked Keiko to present his flukes in the voluntary upside-down position, a routine behavior in his repertoire. As Keiko rolled ventral and drifted to my left, I reached out to grab the leading edge of his massive flukes. This required no small amount of effort even in still waters. Amid the contrasting movement of light chop and the *Draupnir's* opposing shift, the effort required some modification. Robin tied a short loop and put the line around my chest. The added leverage enabled me to lean farther out from the platform. Anchored by the rope, I could use both hands to grab hold of Keiko's right fluke and pull it to the edge of the platform.

With the white underside of Keiko's fluke shining in the sun and nearly in my lap, Robin joined me on the platform blood kit in hand. He is one of the most proficient technicians at taking bloods on marine mammals that I have witnessed. It's a talent that was especially beneficial under the confounding shifts of boat and whale.

Holding the syringe between his teeth, and the butterfly needle in his right hand, Robin steadied himself with his right elbow on Keiko's fluke and his shoulder pressed against mine. With his left hand he took the alcohol swab from Tracy and cleaned the spot where he planned to insert the needle. In the process he located the most conducive vein. Satisfied that he had the ideal spot, he locked his right arm through my left. He then inserted the needle while pressing the ball of his hand against Keiko's fluke to steady his approach. As Robin worked, his concentration was like that of a diamond burglar cracking a safe, feeling for the familiar punch through the exterior wall of the blood vessel. Without fail, Robin hit the blood vessel on the first attempt, evidenced by the crimson fluid shooting up the butterfly extension tube. It was a good "stick." In only a few moments we had ample blood from which we could run multiple tests.

Apart from our exertion in the process, Keiko remained quite relaxed, giving us plenty of time to get what we needed. Robin and Tracy distributed the sample into various vacuum sealed tubes. Michael and I coordinated the boat and Keiko to begin our final approach into Klettsvik Bay. This time *Sili* led the way, first taking her crew to the bay pen and then assuming her position on the barrier net gate. Rather than stop outside the net, we decided to try something different.

Throughout the entire walk, Keiko had remained faithfully beside his ocean-bound escort. Taking advantage of the comfortable position, we simply drove the *Draupnir* directly into the bay enclosure, bounding across the boat gate, located just to the left of our underwater whale gate. Keiko obediently followed, departing from the walk platform in time to submerge and navigate the underwater opening. Once inside the bay, he fluidly rejoined the walk-boat at his designated position by the platform. In practiced form, the crew of the *Sili* responded to Robin's call from the *Draupnir* and closed the barrier net gate. At that, we concluded the first-ever successful open-ocean walk with a trained killer whale. From start to finish, nearly two and a half hours at sea, the excursion went off without a single hitch. Keiko and his walk-boat logged over eight nautical miles on his inaugural outing to the North Atlantic.

Three by Three

With blasting now completed in Heimaey harbor, things somewhat returned to normal, or at least what semblance of normal could be achieved within the Keiko Release Project. The release effort had taken so many turns in recent months there was little consistency in the day-to-day life of the project. Nonetheless, our mandate of release was now more poignant than ever.

Boat walks within the bay continued, pushing Keiko to his limits of speed and distance. If nothing else, our first walk taught us that the conditions of the open sea were very different from those within the bay. Yes, Keiko could ride waves and thus escape the drudgery of swimming headlong into a current, but those opportunities were far and few between in the real world. In the wild, he would have to go where the wild whales went, and they went where the fish could be found. Mother Nature is not so kind as to place their very sustenance at the bottom of a downhill run.

Our extended walks within the bay soon saw a return of the strange, abusive, love/hate relationship between Keiko and his escort boat. He had never even attempted to hit the boat during our trek around Heimaey just days earlier. It seemed obvious that we had unleashed a magnanimous level of new stimulation in Keiko's world and everything in its shadow now fell hopelessly short. We knew instinctively that walks to the North Atlantic must continue. We had exposed Keiko to undiscovered country, the environment that was, after all, the endgame in the progression toward release. We had opened a door and taken a step through that door. Going backwards by continuing to create history within the bay ran counter to every hard-earned step leading up to this pivotal phase.

Ongoing construction in the harbor provided the excuse for more excursions to sea. After the initial blasting, several rounds of pile driving were required to complete the foundation of the new dock. Reverberating jarring sounds and pressure shocks for sustained periods were no less threatening than the opening volley of explosions. Though blunt by comparison to the violence of blasting, persistent exposure to pile driving would be akin to water torture, a slow psychological punishment. Because we had laid the groundwork, approval to continue the walks during the pile driving was swift by comparison. That Keiko returned to the bay reliably in the first walk carried equal weight in securing consent.

Where Away?

Our second journey to sea led us in other directions, with more playful undertones and experiments too enticing not to attempt. Within just one or two nautical miles due east out of Klettsvik, we decided it was time to see how fast Keiko could swim. We had practiced and tested him in short bursts within the bay before, but even the bay wasn't big enough to allow the *Draupnir* to get up on a plane. Here, we had nothing but the beckoning call of the horizon off our bow and thus an unlimited runway. At her best in calm, conducive waters she could muster up about twenty-eight knots with a full crew compliment onboard.

Here we were at sea in the North Atlantic, a killer whale as our playmate. Even the relentlessly safety-conscious Michael was brimming with anticipation. Jen perched in her usual spot above the pilothouse, Robin behind the cabin on the aft deck and Tracy behind my perpetual bouncy ride on the platform, each found something solid and grabbed hold. Keiko, ever dependable, coasted along our starboard side, barely out of my reach.

Steadying myself and wearing the grin of thieves, I nodded at Michael, then quickly turned to give Keiko a prompt with the target pole. Michael punched up the throttle just as I dropped the target back to Tracy and latched myself onto the line holding the platform. Keiko disappeared from sight almost immediately. In seconds, we were on a plane and clocking twenty-six knots, nearly full steam ahead. Having lost visual contact with Keiko, we realized he could not keep up. Yet there was no sign of him in our wake. At best we held our speed for no more than a quarter of a mile, finally giving in that we would have to wait for Keiko to rejoin his escort.

Michael smoothly dropped back on the throttle. The *Draupnir* gracefully slowed and nestled herself down snugly to the waterline. As we came to a stop our moderate wake caught up with us and lifted the stern of the boat in the process. At that same moment, Keiko surfaced alongside the platform announcing his presence with a stiff whale-sized breath.

"Holy crap! He stayed with us!" I exclaimed.

"Whoa . . ." Michael's expression was drawn out with disbelief.

"Did anyone see where he was?" Tracy asked.

Jen answered first, "Nothing from up here. As soon as we picked up I lost him."

"Nobody saw him come up for a breath? That was a pretty decent distance!" I couldn't imagine that he could make that sprint on a single breath.

"He was *right* there," Tracy said, adding, "It's like he stayed right with us. No way he could catch up that fast."

Finally Robin chimed in on the parade of reactions; he was ready to solve the mystery. "Let's go again, this time a little longer."

So we did.

Again Keiko popped up alongside the starboard platform the instant *Draupnir* slowed. We were stunned. Never in our wildest dreams did we imagine that this whale . . . *this* whale . . . could emerge from twenty years in the care of man and magically exalt to the physical prowess of an Olympic athlete. No matter the preparation and exercise of our past many months, a sprint at this pace and distance was impressive by any measure.

Still mystified, we tried again. This time we wanted evidence. In both of the first two dashes, not one of us had seen Keiko. We imagined he must be running deep to a more hydrodynamic position, his body compressed at depth and rifling through the water. At least that much would explain the energetic overcompensation when he surfaced bigger than life at each of the two finish lines.

Eyes wide, alert and fixated on the boundaries of the walkboat, we were hell-bent on being witness to the feat. After all, seeing Keiko move that fast would be a trophy for the ages. This time, I quickly moved from the platform and stood atop the sponson, grasping the top rail of the pilothouse. From here, I had a much better view of Keiko. I expected to track his first movements for some indication of where he made his escape.

Again the *Draupnir's* twin engines pitched up to their drumming howl. Again Keiko immediately vanished amidst the white wash pushed off the side of *Draupnir's* bow.

"Damn it! Where the hell is he going?" I questioned as I kept my eyes glued on our wake.

"I don't see anything," Jen confirmed the same.

But this time we held the pace longer, not by much, but enough. The long shallow swells that day allowed Robin, who often took such chances, to stand on top of the aft engine cowling, nothing to hold or steady himself. From here, he had the perfect view of the *Draupnir's* stern.

Moments later, he exclaimed, "Got him—he's drafting the boat."

"What?" Disbelieving, I scurried back to the stern, crowding the small space with Tracy and Robin, who had now climbed down from his giant surfboard. We all three leaned over the back of the boat and looked straight down. There, only two, maybe three feet below, were Keiko's flukes, or rather, the trailing edges of his flukes, the bright white of each fluke outlined by the black edges. He was upside down, practically plastered to the boat's hull, taking advantage of the liquid vortex beneath the boat that had pulled him along.

"Oh my, God! That's hilarious!" Tracy blurted. "How the hell?"

"Look at that! I can't believe that he can do that!" I was still in disbelief. "Robin he's not even moving his flukes!"

"He's either got a bear hug on the hull, or he's riding her slipstream," he theorized, not knowing which was more likely, perhaps a little of both.

Robin turned and dragging the edge of his flattened palm against his neck, gestured to Michael to cut the engines. Poor Michael, figuratively chained to the wheel, was standing as far outside the cabin door as he dared, longingly stretching to get a glimpse. By the time Michael dropped back to a leisurely trot and finally brought us to a stop, the entire investigation had lasted less than two minutes.

In a repeat performance, Keiko popped up at our side the instant we dropped off plane. Gazing at him now alongside my usual position, he looked almost triumphant. I didn't know how, but he appeared to gloat. His posture? Eyes? The way he tilted his head ever-so-slightly, casting a glance in our direction at each breath? A little of each and more I suppose. But this is exactly how it felt as I stared at Keiko. If I could have placed a caption over his head it would read: "What else you got?"

"You sneaky sucker," I said.

"And here we thought he was faster than the *Draupnir*." Michael had left the helm and was rolling a cigarette, the trademark "knowing" smile pasted on his face.

We drifted for a few minutes, gathering ourselves as the excitement dissipated. We reveled at the ingenuity of the whale. Where had he learned such a thing? The closest resemblance of this behavior had been on the first walk out of Klettsvik when

Keiko repeatedly attempted to draft beneath *Draupnir*. Back then we traveled at no more than four to five knots. More to the point, we had never lost sight of him in those instances. He never got close enough to the hull to plaster his belly into her slipstream. Following some debate, theories, and simple amazed observations, we finally turned to task. We had left the platform in its down position. The entire time Keiko sat patiently at the surface awaiting our next volley.

"Fine, Keiko three: *Draupnir* zero. But he's going to get a workout either way."

Robin answered my challenge. "Position him out with the target?" The question was rhetorical.

"That's what I'm thinking. Keep it short and a little slower until we're sure he's staying out," I said.

"Agreed. Michael, we're going to do a few short approximations, probably only a couple hundred meters and a little slower at first."

"Okee-dokie," Michael acknowledged.

The walk team stayed at sea with Keiko for a while longer, practicing a series of short sprints. Each time we guided Keiko with the target pole ensuring that he would remain to starboard, actually having to do the work on his own. Although these unassisted sprints never matched the previous trials in which we were fooled, Keiko nonetheless impressed us with his swift movement. I believed he could very well have matched the *Draupnir* if he had wanted; however, he could not hold the fast swim for long. At best, his speed came in short bursts.

The Explorer

Our third walk found us eleven nautical miles from shore. We had stumbled onto a nice "flat," a calm convergence of currents that quieted the surface into that of a warm summer lake hundreds of meters in every direction. Similar in ways to Klettsvik Bay on her more forgiving days, birds kept us company both aloft and meandering on the surface. We had started our walk that morning with little direction in mind, our only mandate: the avoidance of other killer whales. In the distance, *Heppin* circled slowly, holding guardianship position over the nucleus of the *Draupnir* and Keiko. Onboard the walk-boat, we waited and watched. Following our run out to sea, and finding an expanse of vacant water (vacant of whales), we intended to rehearse the neutral disposition, affording Keiko his opportunity to explore apart from the dictates of the *Draupnir*.

Stephen Claussen had returned from a short stint at home in the states. Claiming his favorite toy and pastime, he stood at-the-ready by the HDS funnel cannon mounted on the bowsprit of the walk-boat. Stephen was preparing to acknowledge Keiko's first exploits moving away from the *Draupnir* with the now-familiar herring from heaven. Limited by the distance of the HDS, we only wanted one, maybe two opportunities to catch Keiko in the act of moving away. Anything more and we would only teach him to stay within the HDS's range. For what seemed the longest while, Keiko just loitered around the vicinity of the boat. At last, either bored with our quiet boat or distracted in other ways, he slowly moved off our stern. His departure from the *Draupnir* was not decisive; rather, he merely meandered away in a westerly direction, mostly drifting at the surface.

It's hard to say how long it had been. We had no way to influence Keiko's willingness to explore on his own. The best we could offer was the careful directive not to interfere. In each case, the initial effort was to simply become invisible offering no attention, nothing to impede his interests elsewhere, not even the purr of the *Draupnir's* engines. In practicing our ghostly posture, we became somewhat adept at passing the time during these blackouts of competing stimuli. Some of us conversed on a variety of topics, from personal adventures home to how planes could fly. Others occupied themselves with lunch or dinner or whatever excuse could be made to snack. At least one of us would find the most comfortable spot on the pontoon-like sponson and pretend to nap. Although a comfortable place to stretch out, the foam-filled sponson was a precarious perch. One could easily end up jolted awake by tumbling to the wrong side and into the frigid water.

Whatever it was that each of us found to pass the time, we all kept an ear or watchful eye toward Keiko's activity. Jen seldom engaged in the extracurricular activities onboard. She was stubbornly addicted or committed to the gathering of research data. Remaining at her post on the top of the cabin, we often listened to her voice as she recorded the occasional notes into the ethogram microphone. This afforded us the luxury of eating, napping or otherwise goofing off. In this way, some part of our attention was always aware of where Keiko was and what he was doing. It was deep into a cloud of drowsy waiting when finally the monotony was shattered.

"I've lost him," Jen said to everyone and no one in particular.

No one replied, but each dropped what he was doing and started scanning the surface around us.

"*Draupnir—Heppin,*" Michael called casually over the radio, "You guys see Keiko?" Michael always said "Keiko" with his own brand of pronunciation. Keiko sounding more like "Kee-ko," the second consonant as hard as the first.

"*Heppin—Draupnir.* The last we saw he was off your six o'clock about 300 meters."

"Copy that, *Heppin.*"

"Anything on the hydrophone?" Robin asked.

Before Michael could relay the question, Steve Sinelli's voice crackled over the radio. Those of us on deck heard the report from the handheld marine radio placed on the engine cowling. "*Sili—Draupnir.* I heard some vocals a little bit ago . . . not whales I don't think but something else maybe, like a pod of harbor porpoises. Not hearing anything now."

"Would've been nice to know," Jeff said quietly. Only Robin and I were within earshot of the comment. He was wearing his usual comfortable smile, hands jammed in the pockets of his Mustang survival suit.

"I think he's off our nine," Jen offered. By now she had set aside the ethogram equipment and had the VHF tracking antenna out. The antenna was exactly the same as those found on houses back when TV signals were predominantly analog. She held the receiver out in front of her in a horizontal position, slowly sweeping it right to left and back again as she scanned the quadrant from seven to eleven o'clock. The device was very light, made of a larger aluminum post with smaller antennae crossing in a perpendicular fashion. If Keiko surfaced, and if the tag's transmitting antenna was free from the water's surface, and if Jen had the receiver aimed in the same direction at the same moment, she would hear a faint beep through the large padded headphones she now wore.

"Yep, got a signal right off our nine once but seems to be moving east. Second signal was closer to ten o'clock."

"Anything?" I asked Jeff. He stood with both feet on the sponson pressing his back into the pilothouse to steady himself while peering in the advised direction with binoculars.

"No—nothing yet," he drew out the "no" with a questioning tone.

"*Sili—Draupnir,*" came the call over the radio. "We've got a small pod of harbor porpoises moving pretty quick to the north—looks like they came from your area."

"Copy that *Sili*. Any sign of Keiko?"

"Negative, *Draupnir,* not from here. Do you want us to move off and follow the pod?"

"No, tell them to stay put for now." Robin responded without waiting for Michael to relay the question. But Michael didn't have to.

"*Sili—Draupnir.* We've sighted Keiko. He's about a thousand meters behind the pod of porpoises moving in the same direction—seems like he's following the pod."

"Copy that *Sili*. Stand by."

"That's probably why *Sili* isn't hearing the vocals. Betcha they saw Keiko or heard him and went silent trying to sneak around," Jeff suggested.

"Do you think we should call him back?" Stephen asked Robin.

"No, let's see what he does. This is great. This is exactly what we've wanted to see." Thus far, Robin was the only one that didn't show concern at losing visual contact with Keiko.

"First time we've seen him show interest in anything. Pretty encouraging that he's curious enough to check them out." I wanted in on the optimism, although admittedly it was a little odd to let him off the proverbial leash without knowing how far he'd go or if he'd return. This was a rehearsal, not an introduction, and those were not killer whales that he was following.

"But what if he just keeps going?" Stephen continued, as if he had read my mind. "How far do we let him go?"

"We should be careful not to call him back while he's actively showing interest," I offered, mainly to Robin. His confidence became my own.

"Jeff, I think we should wait until *Sili* can confirm that the harbor porpoises have moved away before we try calling him back?" Robin was looking for consensus.

"I agree. I don't want to let him get too far, though. It'll get dicey trying to track him down with only the radio tag."

We couldn't gather satellite data but once every twenty-four hours, even then only by downloading the data on a laptop back at base. At present, the only way to track Keiko was via the intermittent signal of the VHF radio built into the tracking device on Keiko's dorsal.

"I can move us slowly in that direction?" Michael said.

"I don't think we should even start the engines at this point," I interjected, not wanting to break the prime directive. Even though the prospect of Keiko following a pod of harbor porpoise was a far cry from a sustainable release, we couldn't dare risk interfering with the very point of exploration and his display of interest in something apart from us. After all, this was what it was all about. Keiko had found something more appealing than the *Draupnir* in this infinitely foreign environment. We needed the confidence to let it play out. "Freaking out" would be too strong a description of the atmosphere onboard the *Draupnir*, though trepidations abounded in that direction.

Moments turned to minutes, which clicked over into a quarter of an hour.

Robin picked up the handheld. *"Draupnir—Sili."*

"Go ahead, *Draupnir.*"

"Can you still see the pod of dolphins?"

"Negative. They've moved out of our sight. We had them two minutes ago moving in the same direction, but they're too far now."

"What about Keiko?"

"He's south of us about a thousand meters. They were moving too fast for him. He's just kinda hanging in the same area."

"Copy. Thanks."

"We can either recall him or see if he returns on his own," Robin said to the crew.

Always a ready opinion to share, I was the first to offer my views, "I think I'd rather recall him than wait for him to return voluntarily, that is, if we're sure he's not on the porpoises."

No explanation was needed. Robin understood the fine line that lay before us. If we waited for his voluntary return, acknowledged him and continued on, we would encourage future returns to the walk-boat versus that of interest in the new world. Although recalling him now also interrupted in the delicate process, it was the lesser of two evils. Jeff nodded and shrugged agreement. No one else seemed to have an opinion.

"Okay, let's recall."

"*Draupnir—Sili/Heppin*," Michael advised, "Going to recall Keiko."

Two clicks of the radio from each confirmed their receipt of the message.

Stephen dropped the recall transmitter into the water off the stern of the *Draupnir*, closest to Keiko's distant position. We made eye contact, and I nodded.

"Recall," Stephen announced.

We waited.

"*Draupnir—Sili*. Any response?"

"Can't tell, *Draupnir*, we've lost visual," came the reply.

Robin didn't want to chance it. "Try the recall again."

Again Stephen hit the recall tone. Again we waited. It seemed forever scanning for a visual, but it was also a fair distance, and Keiko wasn't known to sprint back to the *Draupnir* at the beckon call of the tone. It would take time for us to become accustomed to how long was too long. This time we didn't need the recall a third time. Keiko finally showed up at the *Draupnir*, having evaded our searching eyes during his commute. He surfaced at the platform on the starboard side without lifting his head. We had been so consistent in delivering his reinforcement on or below the surface that he very seldom lifted his head to acknowledge the world above the surface.

Upon the reunion, confidence among the release team also returned. Conversation bent to the excitement and success of his outward interest in the pod of passing dolphins. Had we let it, the scene might have elicited a search and recovery party among the boats in formation that day. As it was, Robin stifled the knee-jerk need to curtail Keiko's first real adventure, and in so doing, preserved the neutrality of man-made things.

Less Jeff, Robin and myself, the staff's acceptance of the unexpected exercise seemed feigned at best. The idea of release was still so intuitively contrary to the usual mandates of caring for an animal. Nonetheless, at least the two or three of us were genuinely ecstatic at Keiko's display of curiosity. Until now, we had never seen evidence of the extrovert animal we had hoped to discover. Keiko's outward curiosity that day was at least a morsel we could hang our hat on, something akin to the type of animal that could someday find the means to survive on his own. At least that's what we chose to believe, and for the time being, it was a welcome shot in the arm.

In the world of man, the release project floated on a sea of paperwork. As we amassed time at sea, others worked to conclude the final release permit, one that would give us the means to introduce Keiko to his kind. That day was nearing, and everything we did now was a step in the direction of freedom; freedom not from the bay or bay pen, but freedom from the tethered half-release of the walk-boat. Though it was a mechanically brilliant means by which to transport Keiko to sea, we also knew the *Draupnir* and the constancy of our presence with him at sea was a hindrance to his survival. The longer we continued these paired adventures, the more he became dependent on our guidance.

Transference

Bringing together animals that have no history together can be treacherous, even if they are of the same species. In the field of wildlife management, it is understood that social acclimation is a gradual process, especially among highly social animals, such as killer whales. In an ideal setting, the new animal is first introduced to the unfamiliar environment. In Keiko's case this was the open ocean. Once the animal gains experience and familiarity with that environment, he is introduced to the target social group. Again, in a perfect world scenario, the animals are introduced one at a time until they have met each individual in the group. At each introduction, the newbie's presence is paired with something positive, like food. In the simplest form, every time the new guy shows up, good things happen. Over time, the positive association transfers to the new member of the social group.

However, we could not directly reinforce the wild killer whales each time they met Keiko. Though that was precisely what I wanted to do, such blatant influence of the wild animals flew in the face of every ethical consideration in wildlife management. It was impossible to pair a food source with Keiko without also creating an association with man, upsetting the wild animal's natural avoidance of man-made things.

Because we had no control over whether or not Keiko would be accepted by the wild whales, the best we could manage was to reinforce Keiko when the wild whales were nearby. Even so, we had to be careful not to provide much of Keiko's food associated with the *Draupnir* or for returning to the bay. Too much of the former and Keiko would become hopelessly attached to the walkboat. Too much of the latter and we'd end up with a whale expertly trained to return to the bay. The plan was to shift this balance over time and repeated exposure gauging each measured step based on Keiko and his intended family.

While nothing about the plan was ideal, we gambled that repetition would win the war. By providing the bulk of Keiko's daily food requirement each time the wild ones showed up, over and over again, this positive history would eventually "transfer" to the act of staying near the other animals.

In stark contrast, the majority of the FWKF assumed that Keiko "wanted" to be in the wild and would "instinctively" choose his kind. As if a sort of magic, they believed instinct would spontaneously kick-in and override a lifetime of social isolation and human relationships.

The concept of instinct is, at best, obscure and largely misconstrued. The word instinct is often used to describe behavior that we don't fully understand. It connotes a hidden, hardwired ability or skill. In reality, most of what guides an animal's behavior is learned. Shortly after birth, the capacity to learn, and to adapt to an ever-changing world, becomes the dominant factor at the center of survival. Without this important amendment to "instinct," animals or people would never endure beyond their first days or year of life.

Keiko's unusual history with man was the driving force behind his motivation and the choices he would soon make. It was highly improbable that the adopted son would instantly prefer his biological lineage over that of his twenty-year foster family. For all intents and purposes, the wild whales were animals foreign to him.

By design, the walk-boat was at best a temporary step in the path to freedom. It was a one-way means of transport, a prompt that needed to be faded before it became a crutch. The ocean-walk rehearsals were beneficial in exposing Keiko to his new home. But it was time to move beyond this half-step. It was time to begin the social acclimation process.

Events played in our favor. After just three weeks of open ocean rehearsals under the tempered approval, we received the green light to introduce Keiko to wild killer whales. The timing could not have been better. We were willing and Keiko was as ready as he would ever be. Sunday, June 18, 2000, was the day selected for the grand introduction.

As expected, anticipation ran fervently through the hierarchy of the release campaign swelling from Santa Barbara in the east all the way to Iceland. Despite every forensic evaluation of Keiko and the ebb and flow of his tenuous achievements, most considered this day would end in dramatic fulfillment not unlike the sunset scene portrayed in *Free Willy*. The event received the same pomp and circumstance of a presidential inauguration. All we could do was not enough to prevent the circus that converged on the small island of Heimaey. The project moved forward now on its own schedule, shaking loose a series of events set in motion years earlier.

First Contact

A week before the big day, the purposeful introduction of Keiko to wild whales, we met in Jeff's room on the penthouse level of the hotel. The meeting was between Charles, Robin, Jeff, Jen and me. We were deciding the final protocols for introduction: boats and their assignments, proximity of same and the details of each party's responsibility on the water and in the air. This discussion was intended to finalize the step-by-step process as we discussed every angle and possible result.

Most of the plan was agreeable and had already been exchanged in rudimentary form through numerous e-mails among the five of us. The nucleus of the flotilla would be the *Draupnir* and Keiko, supported by *Heppin* to watch our back and prevent third-party vessels from encroaching on the introduction. Two additional boats would be needed, although Robin and I both were reluctant to agree. The *Viking II* would serve to locate and track a wild pod, communicating back to the *Draupnir* so that positions could be coordinated. Yet another vessel would be full of VIPs the FWKF board had to accommodate, a collateral obligation to significant donors. Lastly, the helicopter would be used to film the event and provide additional bird's-eye spotting. But as the discussion continued, other plans emerged that were new to Robin and me.

From day one of our involvement, Robin and I had always viewed Keiko's meeting with his wild brethren as a process, one that would require an unknown quantity of time and repeated exposure. We were very much alone in this perspective. Charles and the FWKF board firmly believed that this was it; that Keiko's first introduction was a one-way ticket and he would not be returning to Klettsvik with the *Draupnir*. When we argued our point, the persistent counterpoint always began with, "Yeah, but what if he leaves . . . ?" or "We need to be prepared." These "eventualities" dictated that the introduction would be treated like a one-off event, requiring full documentation of Keiko's release *just in case*.

Robin and I quickly became distressed by the growing number of vessels and helicopter required to accommodate an entourage of paparazzi, all driven by the notion of Keiko swimming off into the sunset. Our only victory, we refused to allow a diver in the water to film the interaction between Keiko and the wild pod, initially insisted upon by Charles. His agenda largely focused on the creation of an award-studded documentary. After all, film was the lifeblood of Jean-Michel and Ocean Future Society. For our part, we wanted nothing more than what was absolutely required. If left to us, the introduction would consist of no more than the *Draupnir* and one support vessel. Losing that battle, we were both already on edge about every other aspect of the process as it unfolded.

Next was deciding where each boat would be positioned in the flotilla's waterborne configuration and the helicopter's flight path. If we couldn't keep the boats out of the water altogether, we were bound and determined to put them at such distance that they would be rendered just as innocuous. Charles and Jen met our first proposition of several miles with heated resistance. Charles couldn't capture the footage he desperately wanted from such distances. Similarly, Jen could not record the data she needed and which represented the pinnacle of every other data point collected in the path leading up to this day. Jeff supported Jen in her insistence on the data collection, calmly and periodically offering counterpoint. Jen was more outspoken on the issues.

From Robin's and my perspective, there would be many more opportunities for collection of footage and data. The initial exposure to wild whales needed to be positive at best and at worst a relaxed low-key event. To us, this first introduction was nothing more than testing the waters; the outcome of which we would evaluate and refine the process of ongoing encounters. In his own way, Robin tried to suggest as much, but to no avail. The stigma of "release" as a finality stubbornly persisted.

Robin had been on-site for an extended period. Exhausted and irritated from the strife over every decision of late, he became as a pressure cooker reaching its limit. Perhaps a character flaw, Robin would often push himself to extremes during field projects of this nature, willing himself forward on insufficient sleep and the sustenance of Snickers bars, coffee and cigarettes. Although I'd only seen Robin blow his cap on scant few occasions, this was the perfect storm and what ensued would mark an unforgettable and regrettable conclusion to our meeting that night. As the night played out, every frustration Robin locked within himself over the past months surfaced, and all at once.

The layers of this soured onion were slowly peeled back. Soon, it became apparent that Jeff and Jen had been supporting a variety of objectives for the initial introduction, none of which Robin or I viewed as setting Keiko up for success. Those plans were almost exclusively focused on research and to some degree, Charles' independent hopes and desires for unrivaled documentary footage that would put Ocean Futures Society in the history books. It was exactly as if they were changing the game plan, presenting a new playbook at the most critical time in the fourth quarter of a closely matched Superbowl. The realization hit Robin like a ton of bricks. By his assessment, they had intentionally deceived him, withholding information that they knew he would not support and thereby removing him from the decision. Robin took it as a personal offense.

Jen attempted to explain, but she was paraphrasing directly from the mouth of Robin Baird, an orca researcher from the Pacific Northwest, who had zero knowledge of the complex processes at work in Keiko's rehabilitation. Baird had no working understanding of the behavioral protocols, the progress we'd made, or the plan that got us there. It was the same plan that dictated a continuation of precise conditioning steps. Baird wanted genetic data on the wild pod, and Jen was compelled to agree. Among other facts and figures, they wanted to know the lineage of the pod that Keiko would join.

Though the prospect sounded reasonable, genetic data requires tissue, and that tissue is obtained by means of a crossbow mounted biopsy dart. The device would likely render a person hospitalized, but to a killer whale it would be the equivalent of a pinprick. An additional objective involved the use of suction cup tags. These are very basic recorder tags that remain only temporarily on the whales and float to the surface for retrieval after the suction cup loses its grasp.

In either case, the concept of harassing or exciting the wild pod just before they met Keiko was unfathomable. We knew with certainty these greedy demands were going to upset the whole experience for Keiko as well as the wild whales. *Why couldn't they see it?*

Months of pressures and disagreements exorcised themselves before us as Robin reacted to both the perceived deception and the invasive plan in one impassioned explosion. Most of the fallout rained down on Jen. She sat in an overstuffed chair near the east window wearing sweat clothes and socks with her legs folded comfortably beneath her. In this position she had no escape from the torrent.

"You have no idea what you're talking about!" Robin screamed at her, standing so close his accusatory finger was inches from Jen's face. "You call yourself a scientist and say all you want is what's in Keiko's best interest, but you have no idea what's in his best interest!"

Robin's face was beet red. Veins protruded in his neck as he paced away from Jen and then back again. "I'm so sick and tired of hearing this shit about 'what's in the best interest of Keiko' when no one here has any idea what that means!"

Robin never cursed. I had never before witnessed this version of Robin. Only seconds had passed and the rest of us were still trying to reconcile his initial reaction.

"Robin," I said, trying to interject.

He didn't hear me or chose to ignore the call and continued: "I can't believe you're persuaded by this Baird guy who knows nothing about what Keiko's been through or how we're preparing him behaviorally! How in the world can you consider tagging the whales as not harassment right before they meet Keiko!"

Both Jeff and Jen tried to interject but at this point Robin was not listening to anyone. He continued, his words forced through clenched teeth as he glared at Jen. "He is not going to swim off with those whales! You want to turn this into a circus at the most critical time . . . after all we've tried to do here, fighting on every point and every step . . . I'm so sick of your shit!" he spit the words out in utter disgust, every other word almost unrecognizable as they were ejected with such heightened intensity and volume.

He then turned on Jeff who was offering explanation, as much to distract Robin's attention from Jen. Robin momentarily peered at Jeff with a look of disbelief on his reddened face. Nothing Jeff said offered any reprieve. "You guys have been planning this crap for months, and I'm just now hearing about it? Right before we're taking him out there?" He flung his arm out, indicating the ocean in the distance. "This is complete bullshit, Jeff, and you know it!"

At this Robin had his hands out in front of him, palms up. By his tone and his posture he was incredulous, as if unable to accept that anyone could be so deceptive. *That anyone could be so deceptive to him.* Jen tried to respond. But through her tears and Robin's yelling she could not string together a complete sentence.

Behind Robin and to his left, Charles attempted to impose reason. "Robin . . . Robin . . ." he was begging him to stop. Charles' eyes were inflamed and red, the fervid atmosphere overtaking his otherwise stoic composure.

Robin ignored Charles and kept after Jen, the target of his anger. Though Jen's stance in the debate was in stark contrast to mine and Robin's, the resulting attack had far more horsepower behind it than the topics on the surface necessitated. It was indeed the outlet of every frustration, failure, debate and ongoing conflict Robin had buried within himself for a long time leading up to this night. A commonality we all shared; the emotional investment, isolation and pressures conjoined bringing out the best and the worst in us.

Months earlier, the undercurrents were already apparent. Charles sent a note to Robin urging him to seek open dialogue between the four of us. He had suggested that we all four meet in Seattle—outside of the project setting—and come to agreement, or at least mutual respect, of each other's objectives. Due to the demands of the project, the kumbaya meeting never took place. It was unrealistic to expect the four of us to meet off-site. At least two of us needed to be on-site at all times, and over the past five months, that had been mostly Robin and me. In contrast to our schedules, Jeff and Jen had been sent on increasingly frequent expeditions elsewhere on behalf of Ocean Futures. The prolonged separation only served to expand diverging paths between the two duos.

Above all, the nagging insistence of this first introduction as a single "event" only continued to fan the flames of disagreement, which now became a towering inferno.

Robin went the only place left to go—to a personal affront. "You have no idea what real research is and you call this a research project! If any of this crap was out there among colleagues it would be a complete embarrassment!" he said.

Charles, who had been trying to diffuse the situation, now took sharp offense at the personal attack on Jen. He sprang from his chair and was standing almost between Robin and Jen. "That's enough. Robin, you need to stop, walk away. I've had enough—it's enough of this." Charles' words came out cracked and uneven. He furrowed his brow and with that carried off a fairly stern and menacing expression. Both Jeff and I were still caught in the headlights and unable to form adequate responses. It had only been a minute, maybe two. I believe each of us thought Robin just needed to "get it out" and hoped that the conversation would return to civility.

Robin turned and stormed out of the room slamming the door with deafening force as he crossed the hall to our shared suite. I was suddenly aware that the space had become uncomfortably warm. Bewildered, I glanced at Charles and Jeff without an answer to offer. Jeff was first to speak. "Mark, I think you need to calm him down. He needs to calm down . . . or he's going to have a heart attack or something."

"Yeah, I'll talk to him," I said, as I paced toward the door, unsure of what I could do or say.

Charles stopped me, "Wait—just hold on. Give it a minute. I don't think . . ."

But he was cut short as Robin reentered the room much as he had left. This time he spoke with less volume. His rectitude laser-sharp, he had much frustration left to purge. In his brief absence something resonated, some haunting realization that reignited the fire.

"We are not going to make this into a fiasco, I will not allow you to go out there and undo everything we've been trying to accomplish!" Dropping his voice to just above a conversational volume he continued, "Charles, you and I have talked about this; we've talked about keeping this first introduction to a minimum of required personnel and boats. How can you support this?" It wasn't a question as much as it was an accusation. "It's the same bullshit we have had to deal with all along—that Keiko is going to swim off into the sunset! I expect this shit at the board level, but not here."

The breakdown continued at the same intensity for a few moments longer still, defenses and offenses repeating themselves. Finally Robin went back to his room, a phrase of disgust and disappointment left lingering in the air. Stunned, we sat motionless. Jeff turned facing the kitchenette. I had been standing, leaning against the counter the entire time. He glanced at me as he poured a glass of red wine, casting a defeated smile along with a raise of his eyebrows.

After a few moments, Jeff broke the silence in his distinctive soft tone. "Wow! I don't know what to say to that? Is he okay?" The question was addressed to me as if I were Robin's keeper. There was sincerity in Jeff's questioning concern.

"I don't know. I'll talk to him, but I think we should leave him alone for a while." Even I was hesitant to go in the other room.

Among the four of us remaining, the conversation continued where it had started. This time Jeff carried the ball in support of Baird's wish list and in defense of research objectives. I didn't say much. At once I felt any continuation was a betrayal to Robin. In fairness, it would betray my own opinions as well. I had stoked the fires sure enough. In all of our exhaustive discussions, Robin and I played out every scenario we could creatively dream up. With each supposition, I was vehemently opposed to any excess activity surrounding the spotlight event of Keiko's first encounter with wild whales. My representation of the likely outcomes was never candy coated with Robin. Confronted with what he viewed as indirection from those he trusted, and emboldened by my staunch position on the conditioning goals, the disclosures laid in our lap that night—just days before the introduction—were all it took to set him off with such ferocity.

The four of us never truly resolved our differences. The following days came and went with little ongoing exchange. Yet somehow we managed to work together. By and large, our conversations were limited to project needs. The days of personal well wishes and small talk had vanished, especially between Robin and Jen.

Jeff could not state his case then, but had defended his position in support of the tagging and documentation that night with reason. A clear and present fear existed, prompted by agendas and board members bent on twisting Keiko's release to their benefit. He suspected there was premeditated intent to leave Keiko on his own out in the ocean regardless of the initial result. Jeff believed the only insurance that would protect Keiko was complete documentation of the introduction on every conceivable level.

Had the four of us trusted each other, so much would have turned out differently in days and months that followed. We were the front line. In defiance of any misaligned objectives of the board or Charles or anyone in between, to a person and each in our own way, not one of us would have allowed negligence to reach Keiko. Our experiences and our roles were very different, but we all shared a common vision of Keiko as the priority. Nevertheless, we had failed to recognize the one true antagonist in Keiko's venture to freedom. With a long-standing influence over the project, the antagonist would become readily apparent soon enough.

The Big Top

June 18, 2000. At morning's first full light, the harbor was already bustling with activity. Boats and crews were assigned. The meeting of boat captains was under way. The last-minute communication protocols reviewed. Walk formation maps were set out and boat and helicopter positions acknowledged. Everything was carefully and meticulously laid out, checked and triple checked. Equipment was staged on each vessel—from cameras, batteries, and film to the all-important sat-tag, charged and ready for a potentially record-breaking journey across unknown distance and time.

The entire release team was on-site, pushing the envelope on the allowed days in-country. Even the risk of overdrawing on the bank of expatriate days did not matter to project management; they considered the prospect of this being many a person's last rotation a very real consequence of the day's intended activity. Even so, the full compliment of staff proved manageable. At least every one of the regular cast was familiar with the mandates surrounding Keiko. His walk formations now fell into sync with ease. But it wasn't that simple. Heimaey was now host to a variety of newcomers, all there to be a part of history, intent on witnessing the glorious finale to a world-renowned animal welfare event.

On this day, arguably the most delicate of any venture to sea yet, we prepared to escort Keiko to his own kind under the attached scrutiny of five waterborne vessels and one aloft; a veritable floating and flying parade around the waters of Vestmannaeyjar.

Assessing the scene in the harbor, I could almost taste the same feverish hostility that Robin had vented just a handful of days before. *Give an inch, and they'll take a mile*, I had thought. The detailed progression we painstakingly erected and fought to protect was becoming a circus, a show for spectators everywhere I looked. At the back of the line behind every ego and agenda stood Keiko, the featured act.

It was Tom who revealed the first in a series of ruinous impacts that painted the slippery slope already unfolding. During the loading of boats, he had seen crossbows among the assorted equipment stowed onboard the *Viking II*, the tracking vessel charged with locating and identifying the ideal pod for introduction. They intended to either attach suction cup-tags or biopsy dart the wild whales—or both.

Beyond the already overbearing presence of the *Viking II* tailing their every move, the idea of harassing the wild whales with forceful attachments was beyond comprehension. It was the treacherous result of a clear divergence within the project's leadership, an ironic twist of fate.

Jacques Cousteau himself set up many a vivid camera shot that drew the world's attention, cultivating an insatiable fascination with the ocean and her inhabitants, myself included. When I grew older, it saddened me to learn that many of the famous explorations and purportedly candid encounters were staged, orchestrated for film, many times at the expense of the starring animal.

Ocean Futures, the organization on the frontlines of Keiko's release, was a documentary filmmaker as well. By its nature, the organization was predisposed to "getting the shot." So it would be that the very subject of their interest and his needs would likewise be dismissed in the heat of the moment. They protested animal captivity, likening it to abuse and genocide. They saw themselves as great protectors; their mission without reproach, but they were in fact the first and worst offenders.

As we stood ready on the docks of Heimaey's harbor, we knew any chance we had of stopping the intended assault on the wild whales had vanished, both by the hostility of the first discussion and the limited time with which to redress the issue at this late hour. Robin instructed Tom to simply observe and report. He couldn't put the onus on Tom now for an earlier failure to find compromise on the matter. Equally as discouraging, Robin knew this was but another aspect of a suffocating agenda deeply rooted and expanding rapidly. Literally and figuratively, we continued to fall back, retreating to the only vestige we could control: the walk-boat and Keiko himself.

Breaking Dawn

Viking II was required to participate in the fateful introduction. Her role as tracking vessel was to locate and identify the ideal pod that Keiko was to meet. In this capacity she was first to put to sea, several hours in advance of the walk-formation. Captained by Siti, the *Viking II* carried with her Robin Baird, the orca field researcher intending to biopsy the wild whales. Also in her crew were Dr. Lanny Cornell and Tom. It was Tom's duty to provide guidance to Siti in following the protocols specific to the *Viking II* and her station during the encounter.

Our lone eye-in-the-sky, the sleek modern helicopter with call sign *Zero-Nine-Zulu* carried Charles, Jeff and the videographer. Their job was to keep perspective on the field of play from far above. In the final agreed plan, *Zero-Nine-Zulu* had to remain at a minimum of 500 feet off the deck or above the water's surface. *Viking II*, with the assistance of *Zero-Nine-Zulu* would locate a pod of whales, then track that pod communicating their position. Though we knew more would be attempted, *Viking II's* assigned duty was no more than to identify the make-up of the target pod and track their calculated heading, thus allowing the *Draupnir* to assume an accurate position in the path of the wild pod and at the appropriate time.

Shadowing *Viking II* at a minimum of one and a half nautical miles was another borrowed vessel. This was the tour boat *Vikingur*, which had frequented the no-fly zone near the barrier net over the past many months, always playing a game of cat and mouse with our patience.

Tour operators on Heimaey promised their patrons a peek at Keiko, the world's most famous whale. Although we pleaded with the owner to respect the distances we required, he always pushed those boundaries, at times putting his bow almost on top of the barrier net's buoy line, thus giving his guests a front-row view of the release project. The *Vikingur* was our own private paparazzi, repentant after each offense and yet constantly duplicating the affront for the benefit of paying onlookers. Of course we went to great lengths to eliminate any form of unintended enrichment created by the intrusion. But there were times that Keiko would sit just opposite the barrier net, looking up at the touring spectators piled to one side, listing the boat and lending themselves to every form of visual spectacle imaginable for the whale in training.

Now, on the inaugural introduction to wild whales, the pesky *Vikingur* was awarded the lofty position of VIP boat. She carried the bulk of project spectators, favored appointees, agents, regulators and anyone to whom OFS and the FWKF owed favors. To me, she was nothing but another unwanted vessel, tangible evidence of the lack of understanding that ran fathoms deep. What should have been the most minimalist venture to sea was by all accounts drastically upstaged with the presence of man and man-made things.

Completing the flotilla were *Sili* and *Heppin*. In their usual roles, the two smaller vessels would surround the *Draupnir* and Keiko at a half-mile distance. Their duty, as always, was to shield our star from third-party interference. On land, two teams of two traversed the island by vehicle and relayed information when necessary from their land-based viewpoints. Often, the island natives fulfilling these roving positions would be the first to spot nearby pods, as they were well acquainted with the wild whales' habitual routes around the island chain.

Passing the better part of the early morning, the *Draupnir* and *Sili* were tied up to the bay pen while *Heppin* waited patiently in the harbor for her call to action. Time seemed to pass quickly while our minds were full of speculation and constantly interrupted by the tedium of final preparation.

Among those details was the all-important fastening of the satellite tag to Keiko's dorsal fin. Jeff had become the informal master of ceremony in handling the sat-tag. It was his responsibility to ensure that it was mounted properly and that the electronics were active before Keiko could leave the bay enclosure.

Jeff had gone out earlier in the morning onboard *Zero-Nine-Zulu* and successfully located a nearby pod of whales. Once *Viking II* was in position to track the pod, Jeff returned to the bay pen to affix the sat-tag to Keiko's dorsal. *Sili* stood ready to spirit him back to the harbor on his return to the helicopter once we completed the attachment.

Jeff and I called Keiko to the outside of the north pool of the bay pen, the calmest waters that day in the lee of Klettsvik's incoming tide. The procedure was well-practiced. Keiko obligingly floated motionless while we toyed with his dorsal fin and the tag. In only a few short minutes the tag was readied. At that, the last of the morning's details were completed. The team and Keiko were eager to get under way.

Nearly six miles away the *Viking II* trailed in the wake of a wild pod close to the western shore of Heimaey. The animals were moving east-southeast, toward the southern point of the island, toward us.

We had begun the day just after dawn. It seemed like noon, yet it was barely eight in the morning. Any sense of the clock was confused by the early arrival of the sun on our horizon. It was time to get Keiko to sea.

Provocation

Members of our team later filled us in on what was happening from their positions during the debacle. Onboard the *Viking II*, Tom stood at the starboard gunwale near the forward pilothouse. He was taking in the scene as Robin Baird prepared a suction-cup tag. Tom was conflicted. Throughout the morning he watched as Siti was instructed repeatedly to get closer and closer still to the wild whales. He knew they were violating the introduction protocols. He also knew he was powerless to stop them. Lanny did what

Lanny wanted to do, and Robin Baird gave no heed whatsoever to Tom's polite attempts to dissuade the deliberate rebuff. Tom doubted that either of them had even reviewed the introduction plan; much less would they bend to his authority.

Tom was part of my team, the Behavior Team. That association alone rendered him as no more than a nuisance in Lanny's view. More to the point, Lanny likely knew or suspected that we had placed Tom on the *Viking II* for no other reason than to baby-sit the motley crew.

Time and again, Lanny demonstrated little to no regard for what had come before him. It seemed that he cared nothing for the grand plan of introduction and even less for the months of meticulous work implementing that plan. He was fond of mentioning Bobo the pilot whale, a release he himself had orchestrated more than a decade earlier. It was a release for which he alone claimed resounding success; this despite U.S. Navy documents evidencing Bobo's repeated aggression toward divers and the resulting necessity of euthanizing the whale. By his estimation, he was the lone expert on release.

Thus it was, the *Viking II* relentlessly followed on the perimeter of the wild pod of whales, encroaching as close as they could manage and making repeated attempts to place suction cup tags on those within range. They succeeded on two accounts, mounting the tags on a larger female and a younger animal, probably an adolescent, judging by its size. Throughout the morning, when the wild pod surfaced, each animal in the small group took multiple breaths to recharge their oxygen supply. Then the pod "sounded," diving deep, abruptly changing direction and running for a distance before another series of group breaths betrayed their new heading. The *Viking II* continuously adjusted her course, rejoining the pod. Without realizing it, Tom was grinding his teeth. The zigzag pattern and successive sounding of the wild whales demonstrated clear avoidance of the *Viking II* and her pursuit. He couldn't even radio the *Draupnir* to advise her crew of the situation. Lanny kept the only handheld radio on his person. He could use Siti's radio in the pilothouse, but any form of report to the *Draupnir* would incite a riot if it had any meat. Tom did the best he could, waiting and watching.

The pod consisted of at least one unmistakable bull, formidable in size. He easily dwarfed Keiko in physique and prominence, verified by his towering dorsal fin methodically cutting the water as if a giant dark blade. Fulfilling his protective role, the male repeatedly placed himself between the *Viking II* and the rest of his pod. Two or three others appeared to be adult females. At least two were mothers, made abundantly obvious by the very small calves glued to their side and riding in the larger animal's slipstream. The would-be white patches on one of the calves were a dark mottled orange, a pigmentation common only on the most recently birthed. This calf *is no more than a month old at best,* Tom guessed. The troublesome dance between the wild pod and the *Viking II* went on for hours. Their jagged and disjointed course led the entourage in a general eastward heading, toward the southern tip of Heimaey.

From his broad perspective aboard *Zero-Nine-Zulo* Jeff examined the scene below. In company with Charles, he watched as the *Viking II* traveled nearly on top of the wild pod. Looking out over the expanse closing between the *Draupnir* and *Viking II,* he said, "This is not the right group to introduce . . . too many moms and babies."

Although the comment was made over the helicopter's VOX system, Charles stared down at the unfolding introduction without response. Jeff's trepidation still resonating in his mind, his foreboding words faded in the noise of the helicopter as if they had never been given voice.

On the distant side of the island, *Draupnir*, Keiko and those of us on the walk-formation worked our way south paralleling the island's eastern shore. We beat into the small chop and made for the open area at the southern point. Following coordinates relayed from *Zero-Nine-Zulu*, now back in the air, and the occasional sighting report from Siti, we hoped to reach a site in the path of the wild pod and clear of the island. It seemed the ideal spot was in open waters to the south of Heimaey. We didn't know how fast the group of wild whales was moving, and we knew nothing of their aggravated posture. Onboard *Draupnir*, most of our information came from our aerial scout. But at altitude, they could not discern the telltale signs of frustration among the wild ones rising to a threatening crescendo.

Below and on the decks of *Viking II*, the evidence was all but written in stone. To this point, the *Viking II* had trailed the pod of unsuspecting whales for nearly five hours. They were not entirely unaccustomed to curious stalking boats, though this one was overly persistent. On this occasion their wariness intensified not only by the duration of the pursuit and the presence of newly born young, but also by the odd feeling suction cups placed on two of their members and the loud strange bird now frequently passing above.

This was anything but normal.

But their struggle to evade the *Viking II* had been fruitless. Either from helplessness or fatigue or something of both, the pod largely gave way to the needs of the young among them and adopted a slower pace allowing more visits to the surface for air. Hours into the progression, they held a more consistent east-southeast heading. The pod and the *Viking II* were now only two nautical miles apart from Keiko and the walk-formation. In the distance, *Draupnir* and Keiko were just rounding the southern point of the island. The convergence of the two paths became more apparent. The meeting would take place somewhere just south of Heimaey within sight of shore.

The agreed protocol called for *Draupnir* to stop and hold position approximately one nautical mile from the wild pod. This was a programmed pause in the countdown to introduction intended to ensure that both Keiko and the wild whales were aware of each other. It goes without saying that nothing good would come of a surprise encounter. Caution was in order; we had no real way of knowing when either Keiko or the wild whales would each become aware of the other's presence.

Per the staged plan of approach, all boats were scheduled to stop and observe. If nothing obvious presented itself, we could continue forward in half-nautical mile increments, repeating the same check and balance at each interval as the distance between them closed. Somehow that first programmed stop never happened. Without warning we could now see the *Viking II* on the horizon and she us. Only a half-nautical mile remained between.

Radio chatter picked up. The closing gap between the two flotillas shrouded the human orchestra with intoxicating excitement. This was to be expected, but also lent to a quickening where protocol and communication can be altogether lost. Michael's experience and wisdom prompted him to suppress the chatter, and that he did, with measured authority.

"*Draupnirrr—Support,*" he drew out the introductory call. "Ahh, we need to cut the chatter. I don't need to know everything the whales are doin' or what everybody thinks. We need to keep the channel open unless its sumpthin' serious."

It was vital that the *Draupnir* be able to communicate on a moment's notice. No one responded to Michael's reprimand. Nonetheless the supplementary reports quieted for the moment.

At half a nautical mile apart, every vessel in the formation should have been dead in the water with engines shut down. By our estimation, this was more than close enough for the whales to realize each other's presence. On the contrary, that awareness would be unlikely in the midst of rumbling engines and the distracting cavitation of boat props. Steadfast in her leading role, the *Draupnir* shut down. Michael called to all boats to become neutral: to kill their engines and observe.

Aboard the *Viking II*, Lanny pushed Captain Siti to continue onward. Tom debated the issue. Then his mind cleared: this was enough. He would purposefully impose the exacting mandates that all boats assume a neutral position. It was his responsibility. Perhaps he'd had enough after hours of watching what equated to harassment. Perhaps he was emboldened by the visual contact of his compatriots just a few hundred meters distant. Perhaps it was the final sight of a lone female defiantly slapping her flukes on the surface in a display of irritation. Tom himself didn't know, but it didn't matter, enough was enough.

"Siti. Stop the boat. We are supposed to shut down the engines. You need to stop the boat," Tom pleaded and instructed at the same time.

Lanny stood to one side in the small confines of the pilothouse.

"You'll do no such thing. Keep the boat on this heading," Lanny retorted, ignoring Tom.

Siti, so good-natured, did nothing. He looked from Tom to Lanny and back again.

"We're too close. At half a mile we're supposed to shut down . . . all boats are supposed to be neutral," Tom addressed Lanny directly this time.

"Get the f-ck out of here," Lanny snapped at Tom. "We're nowhere close enough yet. You have no idea what you're talking about." To Siti he commanded, "You keep going and don't do anything that I don't tell you to do."

Steaming with anger at the rebuke, Tom left the cabin and moved to the back of the boat's lengthy outside deck. *Life's too short to deal with assholes like this one,* he thought.

For the briefest of moments he considered pulling the ignition key from the helm thus physically preventing Lanny from interfering any further, but only for the briefest of moments. Tom was not a confrontational type. He knew well enough that anything more, whether it be words or action would only erupt into a catastrophic scene at a precarious moment. It wasn't a fear of confrontation that held his tongue. It was most certainly a loathing of everything Lanny represented and how he conducted himself.

The *Viking II* was now only a few hundred meters away. I stood on the platform holding the topline, keeping one eye on Keiko. He sat before me, no sign of anything unusual in his disposition. Outwardly he appeared as if we were on just another leisurely stroll about the ocean.

"Robin, what are they doing?" I asked as I looked back and forth between Keiko and the vicinity of the wild pod, which we could not see. We only knew that they were near the *Viking II*. Oddly, the tracking boat was still drawing nearer.

Robin ignored the question and instead went directly for the solution. "Michael, tell the tracking boat to hold position and shut down her engines."

"*Draupnir—Viking.* Hold position and shut down your engines."

The reply didn't follow the usual acknowledgments. Outside the pilothouse we couldn't hear, but we knew it must have been Lanny on the radio who responded. Michael called out through the open port in the window.

"Robin, they want us to move closer. They're saying the whales are heading off, and we're not in a good position."

"Okay, let's move a bit further in," he replied as he indicated south with his hand held out in a flat pointing position. "Just a few hundred feet, Michael, I don't want to get right on top of them."

I could hear the entire exchange in the background but kept a laser focus on Keiko expecting that some overt sign would be forthcoming, something that would indicate awareness of the wild ones nearby. There was nothing. Keiko seemed to be the most relaxed individual across the entire assortment of species and craft. Still, I did not want to be "holding" Keiko's attention when the discovery occurred.

From the beginning it was mapped out that Keiko's first delicate meeting with his own kind would take place during his emancipation from the *Draupnir*. It was engineered to be a calm and distanced unification. At the very least we wanted to create the ideal conditions for a passive encounter, one propelled by genuine curiosity on the part of both Keiko and the wild whales. It was important that I was not splitting his attention when he awakened to their presence. I could feel the tension in my body. I wanted to get to a stopping point, break from Keiko and get out of the picture. In no time, the expanse between the *Draupnir* and *Viking II* had closed to that of a football field.

"Michael, we need to stop . . . go neutral . . . hold right here." Robin and I must have been on the same wavelength. He voiced exactly what I was feeling. It seemed everything was happening at once.

The *Draupnir* had only been crawling forward. She stopped easily and did not linger. I stepped over the sponson, ready to retract the platform and signal a neutral disposition. In the background my mind recorded sounds of a helicopter, radio chatter, boats . . . more than one, but I was unable to define them with any exacting clarity in my blurred periphery. There was yelling. *Yelling? Why are we yelling?* Communications are almost always over the radio. At sea, even short distances swallow the spoken word.

"Get control of Keiko," Robin barked as he moved to do it himself. I wasn't fully off the platform yet and was able to get there first, dropping the platform hastily in the process.

Struggling to catch up, to gain meaning from the sensory assault, I could make out Keiko in front of me and the *Viking II* off our starboard stern. I was moving to get Keiko to follow. I heard Robin shouting at the *Viking* to shut down. I heard someone urging the *Draupnir* to move Keiko. I heard the engines behind me turning over to start. I heard Tracy's voice. I saw the bow of the *Viking II* first in front of me, coming directly at us, and then turning to her starboard. I saw her port broadside exposed. She was close. She was very close.

The gap between became a narrow alley darkened in the shadow of the tracking boat. *Holy shit, where are the wild whales?* I realized in the middle of the overload that *Viking II* had stayed with the whales just as we had stayed with Keiko. Here she was right on top of us. *What in the hell are they doing? How did they get so close?* Another realization. At once the disorder coalesced to recognition. Somewhere in the middle of this bedlam was a pod of wild killer whales. Keiko sat before me, still willing to hold his faithful position at the extended appendage of the *Draupnir*. He had no clue. We needed to move.

Just as we were gathering ourselves, about to put some distance between the tangle of boats and whales, Keiko plunged explosively to the depths. A forceful exhale burst at the surface sending a disorganized geyser two meters into the air. He moved downward so rapidly that he left a momentary parting of the water where his head had been. Involuntarily I threw my arm up and jerked back away from the fray. The void closed over him in a miniature whirlpool. I had no way of knowing how close the wild whales were or what was happening beneath me.

For all intents and purposes I was largely unprotected. The meager platform on which I sat could easily be sent topsy-turvy if caught in the middle of a whale-sized feud. The lightening speed and forceful movement that would detonate from a charged mix of whales would likely render bystanders as insignificant as a fly hitting a windshield at highway speeds. To my good fortune, both Keiko and the wild whales went elsewhere for the moment, presumably straight down. The more imminent threat came from the *Viking II* who had finally cut her engines and now drifted broadside directly toward the *Draupnir*.

Where Keiko had submerged now became a closing corridor between the boats with only a few feet separating the two hulls. Only seconds had passed. Keiko had gone deep and was nowhere to be seen. In the depths, despite suspending water clarity, all that was visible was a vast array of bubbles thwarting any chance at deciphering the chaos beneath.

In the moments following, exchanges rifled back and forth between the two proximal boats and with others over the radio. Unable or unwilling to feign interest in the premature speculation, most of us locked our gaze outward looking for any sign of Keiko or the wild whales. There wasn't yet time or testimony enough to itemize the jumble or organize thought. Those of us on the frontlines knew without need of scrutiny that the introduction had derailed as if a great and mighty freight train, the cataclysmic wreckage still emerging before us as the smoke settled in the aftermath.

We sat adrift for the longest time. No one knew with any certainty what had transpired in the vast watery space below us or what would come next. We did not know where Keiko had gone. For now, the entourage of boats merely waited for *Zero-Nine-Zulu* to provide some indication of direction, some idea of Keiko's whereabouts. Sightings were confirmed from one of the periphery support vessels: the wild pod had run far off in the distance due south from our position, moving swiftly. All we knew for certain was that Keiko was not with them.

The Unraveling

Two hours after the debacle, *Draupnir* and her crew made a hurried stop in Vestmannaeyjar Harbor. Michael insisted that we top off the gas tanks, knowing that we could be in for a long night. Tracy and Jen disembarked and remained on the island. Brad Hanson came aboard along with the VHF radio tracking equipment. Blair, Michael, Robin and I never left the boat even while she was at dock. The entire exercise took less than twenty minutes. In no time—but all the same an agonizing delay—*Draupnir* and her crew were swiftly back at sea and joined in the search for Keiko.

Zero-Nine-Zulu was in the air scanning the ocean's surface. Onboard *Draupnir* we motored in a general northeasterly heading at about ten knots, waiting for some indication of Keiko's whereabouts. Robin's concentration was intense. He stayed fixed on the bow scanning the horizon, willing his eyes to focus beyond their normal capacity and looking for any telltale glimpse of activity at the surface. Nothing.

It had been several hours since Keiko had contact with the wild pod, and no one had seen him since. We knew his general heading was most likely north-north east of the island based on one unconfirmed aerial sighting immediately following the incident. It didn't much matter; a killer whale can cover a great stretch of the sea in that span of time. There was no telling where Keiko could be, and light was fading fast.

No one onboard uttered a word. Everyone had his eyes on the horizon covering every direction outward from the boat. We were each lost in thought, wondering what Keiko's condition might be and what would happen if or when we found him.

Robin broke the silence calling everyone together at the back of the pilothouse. I had not seen this level of intensity and weightiness from Robin often, but each time I did, it was paired with a profoundly sobering event. Every soul on that boat gave him his undivided attention.

"I intend to get this whale back. I realize that I'm going against some by doing this, but in my opinion this is exactly what's defined in the protocols as an intervention situation."

After a brief pause without breaking eye contact, he continued, "Keiko is not with other whales, and the introduction was obviously traumatic. I'm not asking for your approval. As chief of this boat, I am making the decision to find Keiko and bring him back to the bay. If you don't agree, I will take the *Draupnir* in to harbor, and you're welcome to get off the boat. But I need your decision right now before we get too far from base."

No one said anything, each waiting for another to go first. Then Michael spoke. "Well, I'm in, Robin. I think we need to get him back."

Blair and Brad nodded in agreement. Robin never looked at me; he knew there was no need. It was settled. The *Draupnir* crew would go after Keiko. We would search until we found him . . . and bring Keiko back to the bay enclosure.

Standing around the back deck, we couldn't do anything but speculate. We theorized about what had happened, Keiko's reaction, and the reaction of the wild pod. No one really knew. Although we all had somewhat different vantage points at the time of the event, no one could see anything that went on below the surface.

Nonetheless, we dissected the event over and over, expecting that two divergent observations might combine to provide us some insight as to Keiko's whereabouts. As hard as we tried to anticipate Keiko's actions, we could not. Our discussion did nothing but pass the time.

Hours passed at an agonizing crawl.

All the other boats from the formation were back at dock in Vestmannaeyjar. It was now late afternoon, bordering on evening. Only the helicopter and *Draupnir* remained in the search.

Finally the radio crackled to life. The helicopter had spotted Keiko. Every one of us crammed into and around the outer door to the pilothouse straining to hear. Michael and Robin stood opposite each other with the radio between them looking expectantly at the receiver. Echoing like tin from within the cabin we heard the report.

"*Draupnir*, Zero-Nine-Zulu. Copy."

"Zero-Nine-Zulu, *Draupnir*. Go ahead." Michael was the boat's captain; it was his responsibility to make the reply.

"Take down coordinates six-three degrees two-zero minutes and four-five seconds north, repeat 63-20-45 north and one-nine degrees nine minutes four seconds west, repeat 19-9-4 west. Acknowledge."

"Zero-Nine-Zulu, *Draupnir*. Repeat coordinates 63-20-45 north by 19-9-4 west. Copy."

"Affirmative, *Draupnir*. We have visual contact. We have positive sighting . . . advise heading east northeast, repeatedly circling then continuing course. Be advised fuel is short . . . heading back to base."

"Copy that Zero-Nine-Zulu. We are closing on your location at twenty-six knots. Tracking equipment onboard. *Draupnir* out."

Michael penciled down the latitude and longitude readings and had *Draupnir* punched up to her maximum speed before the radio transmission was complete. We were making a beeline for the helicopter's reported position. Any other time, we would never be able to hold twenty-six knots; however, on this particular evening the North Atlantic tolerated our urgency. There was a following sea, but only long shallow swells, and the *Draupnir* easily outpaced them.

Ship to Shore

On the glassy swells the *Draupnir* rode well, allowing Robin to stay perched on the bow, one leg up on the pontoon to steady himself, his right hand grasping a line tied off to the bow stanchion. There was a tapping on the pilothouse window. Michael motioned for Robin to come back to the cabin.

Back at the hotel the crew overheard the helicopter's report on the base radio. "Robin, Charles on the ship-to-shore wants to talk to you." Michael had to shout through the cutting wind and roar of the engines.

Robin navigated his way around the pilothouse taking care to maintain a grip on the cabin's support rails. At this speed even a small "bump in the road" could throw him from the boat.

Inside the pilothouse Robin took the phone from Michael. "Hello?"

"Hi, Robin, it's Charles. . . . I have Jeff and Lanny here."

"Okay," Robin replied, adding under his breath, "This should be interesting."

Lanny spoke first "What are you guys doing?"

"We're going to get this whale back."

"Why?" Lanny challenged. In his signature condescending tone, he stepped up the attack, "That's against the protocols that we set in place—because he didn't go with those whales doesn't mean he won't inadvertently go with other whales. He may be heading home, and you guys are calling him back. You talk about going against the protocols. Our protocols were always that if he decided to go off on his own to let him go!"

The conflict of interest behind Lanny's motivation was crystal clear to Robin. He muttered, "The bastard just wants his success fee." In the midst of this desperate mess, Robin's patience with him vanished.

"Lanny, you're wrong! That was never the protocol! From day one, in our first meeting at the hostel, we all agreed successful reintroduction would be only in the case of his successful integration with other killer whales. Right now he's alone, he's traumatized, confused, and he doesn't know where he's going!"

Not willing to back down, Lanny pressed, "But we said we wouldn't immediately intervene—that we would allow time to observe his disposition and then make a decision whether to recall him to the boat."

"We are already approaching fifteen miles from the island," Robin snapped back. "If we allow him to go any further away, we will be too far from our base of operations to be able to monitor his disposition and/or intervene should that become necessary. In my opinion, the bottom line is that he is not successfully integrated. The initial introduction was a fiasco, and Keiko is simply running scared! My intention at this point is to find him—make an observation—recall him—and bring him back to Vestmannaeyjar. If the final decision is to allow him to go off on his own, then that decision can be made after we bring him back—and that's a decision that you gentlemen will have to make on your own . . . without me."

Before Lanny could speak again, Charles interjected in a calm reassuring voice: "Robin, we have talked to members of the board, advised them of the situation, they want us to make the decision about what needs to be done. I think Jeff and I agree that you should bring him back—once you locate him—you should bring him back."

Lanny wouldn't let it go, "Well, I think it's wrong, and I disagree."

Only Michael and I overheard the conversation. It didn't matter anyway. The crew was already decided and in unanimous agreement that Keiko could not be left alone. Despite any varying opinions on the actual introduction to the wild pod, it was clear that Keiko was not with the wild whales. Everyone on the project, including Lanny, knew that he would not survive if left on his own.

Recall

It felt like almost an hour before we finally spotted the helicopter; in reality it was probably only minutes. *Zero-Nine-Zulu* passed us overhead in an instant, roughly two-hundred feet off the deck. The helicopter's crew pushed the limits of its fuel reserves and were heading back to base with their own sense of urgency. Another indeterminate while passed as *Draupnir* held her pace relentlessly. Still we saw no sign of Keiko.

According to the reports from the helicopter, Keiko appeared confused or disoriented, swimming with the current for short spells and then turning abruptly and swimming in a small circle. He had continued this pattern for the brief period the helicopter was able to observe him. We could only guess at what this might indicate. At best, he was lost and simply following the prevailing current. At worst he may have sustained an injury from the wild whales or something else we could not have predicted.

Finally, approximately forty minutes after the aerial sighting, we approached the stated coordinates of Keiko's last-known position. Michael slowed the *Draupnir* roughly a quarter nautical mile from the exact location given. Brad climbed atop the pilothouse with the tracking gear. Headphones on, he held the television-like antenna out in front of him and aimed it ahead of the *Draupnir's* bow. He swept the antenna side to side very slowly, listening intently. In order to keep her steady and minimize engine noise, Michael again slowed the *Draupnir*, this time to a crawl.

Again, an eternity seemed to pass. Besides the low rumble of the engines, it was silent as each of us looked at the horizon trying not to blink, then at Brad, searching his expression as he listened for a signal. The sky was darkening and made it highly unlikely that anyone would be able to actually see Keiko, but we stared at the sea just the same and strained our ears for the familiar blow of his exhale we wanted desperately to hear.

The Icelandic summer never really gets completely dark. Instead, evenings become blanketed in a surreal backlight trapping the world in what feels like eternal twilight. Most times it proves to be a curious change from the routine, but this night it only served to augment the already ominous mood onboard *Draupnir*.

Robin motioned me to the bow where he had remained for the last hour. "Put the platform out. I want you to try the recall."

I jumped to the pilothouse and poked my head inside. "Michael, all stop. I'm dropping the platform. Can you kill the engines?"

Michael grimaced and said, "I'd rather not. We're too far out, and if she doesn't start back up . . . " He left the thought unfinished.

"Yeah, bad idea. Just go neutral then."

Without hesitation, I untied the platform and asked Blair to grab the recall tone transmitter. Once the platform was extended I submerged the transmitter. I looked at Robin, and he nodded. I hit the recall tone, holding the button down longer than usual for good measure. We waited.

Atop the pilothouse, Brad was scanning all directions for a signal from Keiko's radio tag.

Nothing.

The surface of the ocean was black, as the evening light had almost completely faded. In the calm swell, we waited for Keiko to break the surface near the platform as he always did in response to the recall tone.

Still nothing.

After ten minutes or so, we began to lose confidence. Surely he must be within hearing distance of the recall? After all, it had not been that long between the helicopter's sighting and our arrival at the coordinates. Michael began motoring at about five knots to prevent drifting in the current. We kept the platform extended, and I remained by the recall.

"I have something!" Brad blurted from the pilothouse roof. "It was faint but in this direction." He was holding the antenna facing due north, one side of the headphones behind one ear, the other he pressed tight against his head listening intently. All eyes turned towards the direction in which Brad steadily held the antenna. In the distance, roughly two miles off, all we could see was Iceland's mainland shore, a faint line on the horizon just slightly lighter in color than the black surface of the ocean. *God, no* . . . Likely we all thought it, but no one dared say it aloud. Michael immediately turned *Draupnir* to port and followed Brad's lead.

The radio tracking tag only worked when the antenna attached to Keiko's dorsal was above the surface of the water. Even if he was close, the signal would be intermittent, and in order to receive it, the antenna had to be pointed almost directly at the radio tag with "line of sight." This meant that the rise and fall of ocean swells could interrupt reception. It was very easy to miss an opportunity. Brad continued to scan in the direction of the last faint beep heard through his headphones.

I couldn't contain my worst fear any longer.

"Robin, do you think he could be on the beach?"

"I don't know . . . I hope not. Try the recall again."

Robin didn't seem to need or want to talk about eventualities or speculation. He was determined to work with whatever we were dealt. At least it was a good sign that Brad wasn't getting a constant signal from the tag. That seemed to indicate that Keiko was unlikely to have beached.

"Michael," I called, "can you turn the platform toward the direction of the signal?"

Michael turned the *Draupnir* to port another ninety degrees and exposed her starboard side to the distant shoreline. I immediately hit the recall tone, this time following it with a slap on the surface of the water with the palm of my hand as hard and flat as I could muster. We waited . . . silently looking and listening, willing Keiko to appear.

Out of the black abyss and with no warning, Keiko surfaced almost within reach of the platform. I will never forget his eyes. His eyes were bugged out of his head; he looked out of his mind. Robin saw it too and quickly reacted.

"Mark, get off the platform. He's wigged out. I wouldn't get too close!" Robin's deft assessment was made all the more dire by the fact that he used my name. He never uses my name.

I didn't argue the point. The sea was an immense black oblivion granting zero visibility beneath the surface. Even above the water everything melted together beyond just a few feet. There was no way I was taking any chances. Keiko was a big marshmallow at heart, but animals and people alike are capable of anything under the right circumstances. We had never seen Keiko like this before, and it was instantly unsettling, if not frightening.

I leapt to my feet and stepped back over the pontoon as the reality of the situation washed over me. *Lanny could give a shit about this animal, his only concern was getting rid of Keiko and claiming a swift and decisive victory.* I turned and faced Robin, who by now had stern look. More prone to reaction than Robin, I let my anger fly unfettered, "I'm sick of this shit. Keiko's totally freaked. And that dumbass claims he's 'swimming home.' For God's sake look at him. He has no clue where he is!"

That moment would become pivotal, intensifying our resolve. Immediately I felt an overwhelming need to protect him—to get the Big Man back to Klettsvik Bay and to confront the decisions that led to this disaster. Every protocol had been broken; protocols that were clearly outlined and all had agreed to; protocols that upheld humane animal treatment far more than the crass dictates of Lanny and those to whom he sold his bag of cheap talk. Lanny had blatantly disregarded everything. He had forced the exposure.

In those moments following recovery, there was no doubt in anyone's mind that Keiko had suffered a tremendous setback. Not only was first contact with his wild brethren an enormously negative experience, it was also clearly evident to everyone onboard *Draupnir* that Keiko was completely disoriented, exhausted and unprepared to make it on his own. As much as I was hurt by the vision of Keiko before me, I was also furious.

Dr. Lanny Cornell had just unraveled everything we spent the last sixteen months working to build. Keiko's first exposure should have been a process—a series of approximations, slow and calm, positive encounters. It should have happened over time and at a pace set by the wild whales and Keiko, not forced on each other in one staged introduction. Again this ill-conceived Hollywood vision of release being a singular "event" derailed the path that might have ensured his survival. This day had put cold human interests laced with ignorance first, and once again the victim was Keiko—always Keiko.

Limping Home

One cannot work with an animal, especially one as exceptional as Keiko and in as surreal a setting, without falling hopelessly in love with that animal. This is not to be confused with possessiveness, which all too often is a hallmark of any animal-related field. As it goes with affection, the singular drive to provide for and protect becomes deeply rooted in every waking thought and action. The distressing venture to locate Keiko, and his condition once reunited with the *Draupnir*, left in its wake a violent storm of emotions for each of us intimately involved in the project.

Forged by the task of release to a foreign wild, our devotion to Keiko manifested itself in the tireless execution of his rehabilitation. If release was the mandate, then we would give him every conceivable advantage to thrive in the North Atlantic. We, a society, brought Keiko among us, and for that he deserved the best; the sharpest minds, and the most attentive to the tasks he had to bear out. And if he could not make his own way, he deserved the careful, thoughtful decision to guide him daily, to provide for his needs without fail.

We gave him a life with mankind and now, the same society to whom his presence had fostered value for his species, sentenced him to the greatest of challenges for his survival. This was a responsibility no one could carry lightly. This plot was human society's own fantasy-based creation, whether or not Keiko was more or less worthy than any other cause did not matter. Here he was and here we were, agendas and politics be damned. Yet, at this acutely vital turning point in the project, human greed and arrogance had unyieldingly levied an immeasurable toll on Keiko's future; a toll that would likely not be undone.

Aftermath

In the moments after Keiko's reunion with the *Draupnir*, everyone on the boat crowded the port beam taking in the scene. Normally the eyes of a killer whale, even those of an alert animal, are a penetrating black orb. But here, the red extents of his eyes were apparent from any position on the boat. I had never witnessed such a stirring disposition in a killer whale. In those moments, I was moved first by relief at finding Keiko, and then by caution at what appeared to be an animal in shock. Both of which quickly gave way to anger coursing through my entire being. This vision before us steeled my resolve producing sharp clarity unlike any other time in the project thus far.

No matter what twisted form of reality was held by Lanny or anyone else for that matter, I decided then and there we would not do this again. My vision was magnified by the intensity of an unblinking stare as I locked my jaw, lost in thought.

Over and over I measured the toll amassed by the blown effort. I imagined the next days, weeks even. *What would be the next steps? How would we resurrect from this disaster?* Walking back through familiar territory I broke the process down to the basics, and then envisioned what the ideal introduction should look like. I toyed with each concept, rolling scenes around in my mind, then reordering them, then changing them again. Finally, clarity settled in my thoughts, and I could see our near future in focus. *Scant few boats on the water, almost a footnote to the presence of whales. That's it. We would need to find a large gathering, a mix of varied pods. A group of that size would most definitely be socializing or in a feeding frenzy . . . not traveling. The presence of a strange whale would mean nothing to them. Once in their vicinity the boats would linger, allow Keiko to experience the sights and sounds. No pressure. He'd have all the time in the world. We would take two boats, that's all we needed. When we find this ideal communion of whales, we'll go neutral and stay neutral, we'll drift as far as the wind or current takes us if we have to. After this first explosive introduction, we'll need lots of repetition, but I'm sure of it—this is how we will go forward. This is how we must repair the damage done here.*

As heartbreaking as the setback was, I had growing determination that we would not be fooled again. This was not the first, but it would be the last time the introduction plan and its prime benefactor would be so blatantly ignored.

"Let's go ahead and put the platform out and see what he does." Robin startled me out of my trance. "If you think he'll do it, I'd like to take a close look at him." He was referring to a body exam. Robin wanted to look him over to check for injury or hot spots that might indicate severe bruising.

Keiko was floating so close to the *Draupnir* I had to ease the platform down, allowing the line to slip ever so slowly through my grip as I lowered the drawbridge-like appendage to the water. Keiko reluctantly moved ducking beneath the black surface, adjusting and then resuming position in front of the platform. Thus far, he had yet to lift his head above the surface, the customary and familiar acknowledgment typical in human-whale relationships. His breathing was frequent, but no longer erratic. This and his increasing stillness calmed my nerves enough that I stepped back onto the platform. Despite Keiko's reputation as an unusually docile animal, I remained ready to spring back, subconsciously assessing the less obvious traits of an animal I did not recognize.

I extended my arm, distally giving the signal for a ventral roll and fluke present (also the safest position in which to protect myself). Keiko took a breath and very slowly complied, moving to my left and toward the bow. Rolling ventral and exposing his white belly, he continued to drift out of the behavior well beyond my reach. Keiko only remained in a ventral position for a few moments, rolling back to a dorsal drift and moving away from the starboard beam of the *Draupnir*. A complete divergence from the whale we knew so well, he was unwilling or unable to acquiesce to my request. In a long slow arch, he returned to the previous position in front of me. The sluggish circle took more than a minute. We tried once or twice more, but each time it was the same. Robin finally said he'd seen enough, taking survey during the slow swim-by of each attempt. It was time to start our trek back to the base of operations in Klettsvik.

The water's surface was abnormally glass-like; the blackness of depth undulating with long slow swells, rising and then passing beneath us in drawn out intervals. I remained on the platform giving Keiko a point of focus as Michael dropped the engines into gear and started to edge the *Draupnir* forward. Keiko took a full breath, pausing between the exhale and voluminous inhale. He began to follow his escort as he had so many times before. However, this time it was different. This time, after only a few meters, he dropped back behind the *Draupnir*, finally coming to a stop and floating at the surface.

"Robin." I didn't need to say more.

"Michael, all-stop," Robin called to the pilothouse.

We held our position. Our stopping prompted Keiko. He made his way back to the platform. Still he did not lift his head, only continuing forward in a restful position at the surface, the unusually calm sea afforded him the opportunity to relax much the same.

We tried again, motoring forward at no more than two knots. This time he stayed with us a little longer. After maybe a hundred meters he began trailing off to aft once more. We could almost see the damp air filling the space between the boat and the retreating whale. Michael slowed but did not stop. At this, Keiko gained ground on the *Draupnir*, finally making his way back to the usual position off the starboard platform. We held this pace and continued the journey back to the bay.

But it was not to be. Once again Keiko began slipping behind. Obvious that he was not able or willing to stay with us, we again stopped the boat. Thus far, we had covered less than a quarter mile.

"I think we're just going to have to let him rest a while," I said.

Robin had been at my left shoulder, just inside the sponson and behind the platform. He didn't respond but rather chewed the inside of his bottom lip in thought, staring at Keiko.

"What do you think?" Michael had stepped outside the pilothouse and was standing to my right opposite Robin.

"I don't know, but we can't keep going. Let's just give him an hour or so and see what that does."

Robin sounded as if he was talking to himself more so than anyone in particular. He avoided using the word "rest" from my proffered assessment, not yet willing to accept the diagnosis. But for the time being, we would stay put just the same. Keiko gave us no other option.

So it was; the five of us got ourselves situated, looking for pseudo-comfortable nooks and crannies around the boat to wedge ourselves into; maybe even catch some shut-eye if we were lucky. The night sky only faintly lighter than the ocean's surface, we could see well enough on the boat, though distant forms blended into dark outlines framing the watery world on which we waited. Keiko remained faithfully at his surface resting position just off the starboard aft beam. Beyond his periodic, whale-sized breaths, all was eerily silent.

Not yet able to consider sleep or any attempt thereof, both Michael and I began messing about with the coffeepot we had commandeered from the hotel on our brief crew change back at base earlier in the evening. There are likely few cases in the annals of history whereby the attainment of a single cup of coffee was so obstinately pursued. By the time our makeshift rig reluctantly surrendered its first cup of black gold, we had taken apart the entire guts of the *Draupnir's* pilothouse. Upending the floor, pulling batteries and creating an improvised converter, we powered up the coffeepot and waited eagerly as our contraption heated up and spread the calming aroma of hot fresh java throughout the small cabin. If they hadn't known us better, our three mates would have thought we'd lost our minds; though no one shunned the communion when offered. It was probably no more than the commodity of hot coffee under such spartan conditions that inspired our streak of ingenuity. Regardless of purpose, it passed the time.

One hour flattened into the next. Eventually it was more tomorrow than yesterday, a new dawn rising in the distance. As shadows faded, fatigue gave way to restlessness. The fog of early morning without sleep made everything seem more than real; gravity heavier, air thicker and thoughts harder to hold onto. Wit no longer colored what remained of our conversations onboard. We were all business. It was time to make another push to home base.

I approached the platform flatfooted and grudgingly. Straddling the *Draupnir's* orange pontoon I looked for a comfortable middleground. There was no certainty that Keiko was going to do any better this time, despite several hours of rest gracefully permitted by the lingering calmness of the North Atlantic. I knew before we started that it was going to be a long slow ride and mine was not a restful place for the journey.

Morning was marked by the arrival of the sun peeking above the horizon. But progress was insufferably drawn out, each mile accompanied with the uncertainty of how long Keiko could persist. Throughout the wee hours of morning and into the day we crept across the watery landscape covering half the distance, almost fifteen nautical miles. The trek thus far had taken half as many hours. At intervals we made attempt to build to the normal walk pace of four or five knots. In each case Keiko eventually lingered back forcing us to drop once again to barely more than an idling crawl. Thus far the only benefit was that we did not altogether stop. So long as we stayed just above idle, Keiko remained reliably by our side.

Ignorance Suffered

Nearing the afternoon almost thirty-six hours after the botched introduction, we could finally clearly see the detailed outlines of Heimaey directly off our bow. Almost there. During the drawn-out walk, Robin had periodically relieved me on the platform, allowing me to stretch my legs or lie flat-out on the engine cowling for a spell. It was during one of these breaks that the radio crackled to life.

"*Viking II—Draupnir.*" The familiar jarring English of Siti's voice made the request.

"Yeaaah, Siti. Go ahead," Michael offered in his usual drawl, an uptick in the middle of "yeah."

"What is your position?" Siti continued.

"Ahhh . . . we're about four miles due east-southeast of Klettsvik," Michael replied as he studied the radar screen for more detail. Then added, "Moving slow. What'cha got?" Michael knew something was afoot. That *Viking II* was back at sea was not lost on him.

"We have whales. Can you come?" came Siti's broken answer.

At this I sat up from the cowling. "What . . . where are they?"

Robin too had turned and was more on the boat than the platform, listening to the exchange.

"*Draupnir—Viking II.* What's your location?" Michael relayed my question, although mine had been more a rhetorical statement than any real desire to know.

"Two miles from Heimaey, south."

"They're on more whales, about two or three miles from us," Michael explained. Using radar, he had guessed their precise location.

"Robin, we're not doing another introduction now are we?" I pleaded more than asked.

"No, we're going back to the bay. Michael tell them we're staying on course for the bay."

Michael relayed the information, but rather than silent acceptance, an unfamiliar voice came across the radio.

"We have a pod of mostly females here, and they seem to be foraging. This would be a good group to bring Keiko to," the voice said.

I grabbed the handheld radio from its charging cradle. I was standing in the cabin door, between Michael and Robin.

"*Viking*, this is Mark. Keiko can barely keep pace with the *Draupnir* as it stands. He's completely worn out." I assumed the education would be all that was needed.

"I think you need to bring Keiko this way. This is a good group ... much better makeup than the whales yesterday. This is a good group for introduction, and we're not that far."

Giving the benefit of the doubt, I repeated the description, "Keiko is in no condition for another introduction. We were lucky to get him back this far."

The voice became more insistent. "He can make it. We're only just south of the island." The anonymous voice never stated his name, but I knew it must be Robin Baird, because it wasn't anyone I recognized from the release team.

To Robin onboard *Draupnir* I pleaded for support. "Robin, there's no way! There's no way we can take Keiko into another pod ... talk about setting him up for failure!" Lack of sleep and the freshness of Keiko's traumatic first exposure was all it took to revive my anger.

Robin nodded agreement.

"Negative *Viking,* we're heading to base," the response was as sharp as I intended. "Keiko is in no condition for another introduction."

At that I set the radio down and left the cabin, not waiting for the response or any form of acceptance. There was a reply, but I didn't listen to the details. This time Michael answered, adding his own descriptions to the state of affairs.

Neither Robin nor I cared to engage the discussion further. The prospect of taking Keiko back to another pod in his current condition was baseless, absurd, and impossible, really. Their ignorance was to be expected. Those in charge had not seen Keiko's condition, nor did they know how disoriented he had appeared just twelve hours earlier. It was the intransigence that was most offensive, that they debated our explanation on the subject as if the decision belonged to the voice on the radio.

The insistent call bordered on an order, arrogantly snubbing our firsthand accounting much as the introduction plan itself was so brazen. Never again. I locked my jaw in defiance as I heard the exchange, now muffled inside the pilothouse, finally come to a close.

Who was this jackass anyway, who clearly believed taking a fatigued and potentially traumatized animal into a new pod would be humane, logical or even possible? Not one ounce of investigation into the previous day's events had been allowed to surface. There were too many people involved, each with a personal agenda, and too few of whom based their trajectory on the needs of Keiko.

The Decision

The sun still high on the horizon gave the illusion of early afternoon, though it was officially evening by the time we made our way into Klettsvik Bay. It seemed the longest day we had suffered in recent memory. Thankfully, Keiko followed through the barrier net without incident. There was no effort to hide our relief when he surfaced inside the bay after *Draupnir* had crossed the boat gate in her customary leading way. To a man and whale, we were all exhausted.

Following our arrival, we tied the *Draupnir* up to the bay pen and off-loaded some small gear while waiting for the night shift to arrive. It seemed unusual for the pen to be so void of activity. For the first time since the whale's arrival to Iceland, the bay had been emptied of her famous inhabitant for very near two days; there had been no need for staffing the pen in Keiko's absence.

Closing out the unexpected journey was a blur as we made our way to the harbor and finally back to the comforts of the hotel. Charles met us at the docks and advised that the staff would be meeting in the solarium. Recognizing the worn-out state of the crew, he offered that the meeting could wait; that we get some hot food first. Dog tired and still engrossed in the unending fiasco, none of us realized that we hadn't eaten in two days. Hunger overcame us at the suggestion.

I was scarcely conscious of the brief reunion at the hotel or changing from our splash suits and wet gear. What resonated was our discussion with Charles as we walked to Café Maria, the town's only pizza joint.

During the short hike, Charles mostly inquired about Keiko, what we thought of the introduction, how he looked after we found him, and what had transpired on the return walk. Robin provided the only narrative necessary. Although I seldom lacked interest in offering my own observations, weariness coupled with acute alignment with Robin's descriptions stilled my urge to speak. Blanketed in the comfortable fog of fatigue, I merely listened. I watched Charles intently. He offered no indication of a direction moving forward, nor did he allow any discourse on the topic. That aspect, in and of itself, was unnerving; so uncharacteristic of the relationship between the three of us.

Charles did not stay with us at the café. He seemed aware of our need to recharge and left us temporarily free from the heavier topic. Any form of evaluation or decisions left to be made would be addressed at the staff meeting upon our return, though we never made it to the fateful meeting.

On our return to the hotel, Charles met us on the penthouse level in our shared room. With razor-like precision, the pinnacle of the issue was laid before us. The meeting had already begun. It was apparent from the start that some direction was already set in motion. They, with no real definition offered on exactly who "they" were, intended to continue the introductions as originally planned, with the full complement of boats and helicopter, tracking and tagging. Everything would be replicated. We were dumbfounded, frozen in disbelief. In a million years we had not anticipated a return to the same mistakes. We assumed that everyone recognized the need to modify the plan. We expected some debate on what those changes would be, but not this, not a complete reprise of the same mangled and reckless approach.

"Well, that's disappointing. You're telling me that no one recognizes the fiasco this introduction turned into?" Robin was clearly irritated, but he never raised his voice. Instead his calmness carried with it a grave overtone that commanded attention at every spoken word.

"No one's suggesting that we don't need to change the approach. I'm saying that the protocols will be followed. We all recognize what happened out there, and we cannot allow that to happen again." Charles was sure of himself.

"We can't go back out there on another introduction with all those boats and all the activity leading up to it. It has to be controlled. We have to minimize the number of boats and personnel involved," said Robin.

"Robin, I understand why you're saying that, but you have to remember that we agreed to this process for specific reasons. You agreed to the approach. I know that protocol was not followed. That's all we're saying, that we keep the same formation, but follow the protocols as they were outlined."

"Charles, there's no way we can have that many boats and a helicopter out there harassing the wild whales and causing who-knows-what kinda distraction to Keiko. It's too late now. We've already created a situation that's probably negative to Keiko anyway." Robin touched on the heart of the issue, but didn't satisfy my obsessive need for behavioral detail.

Interjecting, I expounded on Robin's point. "Charles, we need the introductions to be as low-key as humanly possible, especially now. We've just created a negative association for both Keiko and at least the one pod of whales. There's no question in my mind that introduction was traumatic for Keiko. We can't force his acceptance. This has to be a process, and it has to be passive on our part."

I was repeating what we had argued in heated debate long before the first introduction, but this time we had the advantage of outcome on our side. Keiko did not swim off into the sunset. The Hollywood vision of what release looked like had been vaporized. I hoped that our earlier premonitions would now gain footing where before they had failed.

"Then what are you proposing we do?" Charles opened the door and Robin stepped decidedly through.

"We take him out with one or two boats. That's it. I know the board won't like it, but that's what it's got to be. This is going to be a series of introductions, a slow process that could take several seasons."

Charles wasn't happy with the response or Robin's conviction. After an extended pause, he replied, "Okay, let me talk to the team—get some thoughts and feedback."

Charles' words indicated a possible compromise, but his tone said otherwise. He seemed defeated. He left the room and headed back down to the solarium.

Tom had joined us just before Charles' departure. We summarized what parts of the conversation Tom had missed. The three of us continued debating the finer points of our insistence at minimizing the formation involved in the next series of introductions.

For the present, the issue was eliminating the extraneous man-made presence during the next introduction, creating as calm and nonthreatening an environment for Keiko as we could practically achieve. As importantly, nothing could be done to unsettle the prospective pod targeted for introduction prior to Keiko's arrival. That meant we needed nothing more than the *Draupnir*, the *Heppin* on perimeter guard and one small tracking vessel or the helicopter to spot the wild whales. That was it. Anything more was not practical to maintain over a long-haul and would only excite the environment as had already been proven.

It wasn't thirty minutes before Charles returned, and the conversation continued. For the better part of two hours it went on in much the same fashion: we, unwavering in our demands, Charles unrelenting in his assertion that "they" would not agree.

In the last short absence afforded by Charles' back-and-forth, Robin, Tom, Kelly and I deliberated, hoping to find a possible angle of compromise. Then we reached unanimous agreement. We would not continue to participate if they insisted on maintaining the heavy-handed approach to the introductions. Our ultimatum was absolute. Robin carried the flag when Charles returned.

Charles spoke first. "We'll follow the protocols exactly as they have been laid out, but the consensus is to continue the introduction as it is." He communicated the verdict as if passed down from a higher power.

For a brief moment there was silence. Then Robin responded, his words now softened, though not apologetic. "Charles, you can't ask a man to do a thing that he knows is not right. I will not take part in what I believe to be negligent or that's not in Keiko's best interest."

"I'm sorry to hear that. I don't agree that it's compromising, but I respect your position, Robin. Is there any chance that Tom or Kelly is willing to stay?" He didn't include me in the offer. My involvement thus far placed me clearly by Robin's side.

"You're welcome to ask them directly, Charles, but we've all discussed it and I know they are in agreement with Mark and me."

"Okay."

Charles expression was hardened and serious, but his eyes showed compassion.

His one word acceptance carried with it a timeless weight reflecting many hard-fought battles we had shared over the past many months. The outcome resulting in our departure from the project was surreal. Silence filled the room as Charles descended the stairs for the last time.

I couldn't think of Keiko. The wound just inflicted was far too fresh. Every thought was a struggle to form. Each attempt to make sense of their decision evaporated and became something else entirely. Aside from murky daze, what I knew above all else was that we had to hold our ground. We had never deviated from our insistence that this was a process, even though we had compromised greatly at times. There was no more room for compromise. Not when it came to Keiko. We had given up enough already, and that which we allowed had changed everything. Never again.

That night, I had hoped that the finality of our commitment would resonate, that somehow the morning would bring a new alternative or willingness to consider our proposed modifications. But nothing more came of it. Early the next day, the remaining crew departed for the harbor just as we left for the small island airport and our final journey home from the Land of Fire and Ice.

Reflection

In the years following, I reconnected with most of the staff of the release team. Regardless of the treacherous footing we often found ourselves upon during the project, the experiences and the trials forged lifelong friendships not so easily undone. No matter our diverse backgrounds or what our experiences taught us, we each committed ourselves fully and for that cared deeply for Keiko, his plight and one another. In our own ways, we all wanted the best for the Big Man. That shared vision ultimately bound us in ways we were scarcely aware of at the time.

Those I knew and worked with, and others who were yet to come were eventually exposed to ultimatums levied by the project. Some manufactured by the organizations in charge, and others that slowly emerged from the fog created by passionate commitment to an animal. As the project continued and even years beyond its conclusion, I maintained communication with many of those who dedicated an important period of their lives to Keiko. It is from firsthand testimony of those who carried on that Keiko's story becomes complete.

Keiko in the Wild

The morning of July 20, 2000, the portion of the team consisting of Robin, myself, Tom and Kelly departed the island. At the same time, the remaining staff prepared to take Keiko out to sea once again. It was summer and not only was it a time of high-value weather conducive to the task, it was also the only season in which the wild whales frequented the area. A day could not be wasted.

As the captains each prepared the *Draupnir*, *Viking II* and supporting vessels, the Behavior Team on the bay pen was just realizing that Keiko would be unable to endure another introduction so soon. Physically depleted, his lethargy was immediately apparent and stifled any plans to guide him once again to a pod of wild whales. It was three more days before the team was able to get Keiko to the vicinity of his wild brethren.

Nonetheless the remaining release team eventually took Keiko to sea and to more whales. Between July and October of 2000, the team continued numerous tours surrounding the island chains of Vestmannaeyjar. During these walks they encountered many more and varied pods of wild killer whales.

The expeditions were led by Jeff and Jen, following the established protocols from the original plan. In each chance encounter with conspecifics, the *Draupnir* led Keiko to a position in the path of the intended family, then released Keiko from his position at the side of the vessel and assumed the practiced neutral stance. Following the fateful first exchange, Keiko continued to exhibit clear avoidance of the wild whales, clinging to the immediate vicinity and safe haven of the *Draupnir's* company.

Keiko's reactions to the wild whales on these particular outings evidenced an underlying trauma resulting from the first botched introduction. In both animals and humans, learning that occurs in conjunction with adrenal activity becomes almost "hardwired." Under the stress of aversive conditions, adrenaline coursing through the bloodstream and chemical cocktails taking place in the brain, our predetermined design for survival locks these events firmly in place lest we not forget the circumstances that may threaten life itself. So it went with Keiko's first foray among his kind, the learning taking place and the association with wild killer whales occurred under supremely distressed conditions, the results of which became a defining factor in Keiko's quest for freedom.

Animal Magnetism

That the *Draupnir* represented a calculated risk in Keiko's indoctrination to sea was, by and large, grossly underestimated. We often argued that she must play a static role. By design the *Draupnir* was indeed Keiko's means to gain open water access. As such, she was an increasingly magnetic force in Keiko's life. Robin and I insisted that every voyage to sea alongside the *Draupnir* would create an association, if not a dependence, counterproductive to the aim of social integration. We had expected that introductions to the wild whales would take time and repeated rehearsals. We knew that the *Draupnir's* life cycle in the release progression must be limited and that those limits would be dictated by how and when Keiko's interest in his wild cousins evolved.

It is of vital importance to recognize that food, although instrumental, is not by itself an all-powerful force capable of trumping other and varied forms of positive stimulation. In this regard, the most dangerous form of influence shaping Keiko's choices was people; the enriching human companionship that he had known all his life. At this intersection in the project, the value of human interaction had vastly increased by the sheer purposeful deprivation of same throughout the mean season. Any occasion where Keiko now gained the reward of acknowledgment from his past family was intoxicating to him.

But the notion that Keiko and the wild whales would naturally take to one another proved too difficult to root out. The belief that Keiko would choose his wild counterparts over his treasured escort maligned priorities. As a result, the counterpoising effects of the *Draupnir* were unwittingly increased.

By mid-July, less than a month following our departure, additional activities were assigned to the walk-boat during ventures to sea. For starters, interest in dive data on Keiko led to the novel practice of pointing him down to a submerged buoy that was set at varying depths. To ensure that Keiko would in fact touch the deep target, a camera was mounted on his back by a suction cup apparatus. Via the camera, the training staff could confirm that Keiko had touched the target before acknowledging a correct response and providing reinforcement. Herein, contradictions existed on many levels, not the least of which was the positive experience of learning a new behavior in association with the *Draupnir*.

Deep-dive training was not the only interference introduced during the first season at sea. For reasons not entirely known or worthy of analysis, Keiko returned to the destructive behavior of hitting the walk-boat. Again and again contacting the boat's hull, the satellite tag affixed to Keiko began to damage his dorsal fin where the pins penetrated through tissue. In an effort to minimize the resulting damage, the original titanium pins were replaced with more flexible nylon ones. Against the beating, the weaker nylon pins didn't last and the tag was occasionally found flopping backward after the forward pin snapped in half. In numerous instances, lacking any other practical means, the staff got into the water with Keiko while at sea in order to patch the expensive tag. Getting into the water directly with Keiko in the context of the open ocean broke yet another seal, transferring a long history of human water play to the new world.

Still more alluring research opportunities increased the use of the suction-cup camera, gathering visual data on Keiko's pursuits amid explorations away from the walk-boat. All of these things transferred Keiko's old-world with humans to his new environment. The walk-boat became a veritable center of stimulation and a source of continued reinforcement in the very environment where release would require exacting elimination of man and man-made things.

Jeff and Jen had long maintained a persuasive research orientation toward every decision encompassing their involvement in the release. During our term on the project, the often times conflicting emphasis between research and behavioral necessities offset one another. The foursome worked. Where data was required, Robin and I provided the means to achieve collection in line with strict guidelines shielding Keiko from any damaging influence. Lacking this check and balance, what had previously become a well-orchestrated decision matrix now teetered precariously away from the foundational importance of behavior in Keiko's journey to independence. Bit by bit, albeit unintentionally, the *Draupnir* became a formidable obstruction in Keiko's choice between two worlds.

The team continued on this course of introduction throughout the remainder of the summer season, taking advantage of each nearby presence of whales posthaste. By October, sightings of whales dwindled and opportunities became scarce. Coupled with inhospitable weather typical in the later months, operations on the high seas had to be suspended for the winter. Between late October and until approximately May, Keiko remained in his static bay enclosure.

Throughout the winter the staff maintained much the same program routines as established during the initial rehabilitation leading up to ocean walks. Days consisted of exercise sessions, required husbandry and walk rehearsals within the confines of Klettsvik Bay. But there were differences, and those differences clashed with the long-term goals of release. Preexisting directives we had fashioned to reduce human contact while not discarded, were greatly relaxed.

Among the changes in continued and varied human relationships was a return to playful waterwork, enriching both staff members and Keiko during the monotonous long dark days of winter. Although the impact of these practices cannot be accurately measured, the activities would nonetheless levy a toll to be paid later.

Much like a drug addict falling off the wagon, the pendulum swing of a return to increasing human interaction and activity only served to strengthen Keiko's lifelong reliance on man.

A foundational shift in the project now laid to waste all prior designs. Expressly, the return to Klettsvik Bay and all that was required to keep a solitary whale healthy for a prolonged winter of confinement. Though the prevailing idea of release focused on a single season, multiple years were later considered and always included the prospect of relocating our base of operations. In stark contrast to the shipping channel of Klettsvik Bay, the prospect involved a location in which Keiko could be granted consistent ocean access without the need (and pairing) of a walk-boat.

The back and forth of ocean access followed by confinement for the winter placed unrealistic demands on a staff ill-equipped to navigate such a jarring juxtaposition in Keiko's life. Each season that Keiko failed to integrate with his own kind was met with a return to the known; a return to the security of familiar things void of challenge. If the qualities of an explorer can be isolated, brought forth as a way of life, it is without question that this fluctuation between worlds would not constitute the means by which such a lofty goal could be achieved with any hope of permanence.

Jimmy

As they customarily did, Jeff and Jen headed stateside for a few months during the ocean walk blackout season, leaving a skeleton crew behind to see Keiko through the winter. This time they left a new addition to the team. Jim Horton had well over fifteen years' experience in zoological care. A former SeaWorld employee, he had been a respected senior among his colleagues in the Animal Care Department. Jim had nearly seen it all. He was an ideal candidate brought into the mix at a time that presented specific challenges to sustaining Keiko's health and well-being, an area of expertise right up Jim's alley.

It wasn't long before Jim's talents were called upon. Keiko, as he had done before, began to slow. None of the usual interactions or attempts to stimulate interest bore fruit. Where Jim recognized the telltale signs of a system under duress, countless others might have missed the more subtle clues or mistaken them for simple disinterest or lethargy. He knew this animal's immune system was fighting. It wasn't Jim's experience with Keiko in particular that allowed his timely diagnosis. In fact, knowledge of the historically lazy whale might have only dissuaded his better judgment. Instead Jim saw only physiological struggle.

A lifetime of trusting his gut taught Jim that hesitation can mean the difference between life and death. There are those who would scowl at such a dramatic claim, yet Keiko's past demonstrated that the condition was indeed life threatening. His swift assessment and action were instrumental in saving Keiko's life. Immediate clinical samples were taken and revealed an elevated white blood cell count. Early detection and rapid treatment kept what might outwardly appear a common cold from prospering into something that could overwhelm Keiko's immune system.

By the end of Jim's first winter on the project, Keiko had suffered two distinct bouts of illness. He learned better than to regard Keiko's health casually. Sharing a nearly identical background with Robin, Jim was not a trained veterinarian, but he had treated more fragile survivors than ten vets combined. His analysis was not educated in any lab; it was the product of intuition gained along the hardened road of many rescues and rehabilitations, successes and failures.

Jim believed that Keiko suffered a chronic respiratory infection. Not unlike walking pneumonia, the condition was likely concealed during times of peak activity when Keiko would clear his lungs through the exertion of ocean walks and fighting currents. Back in the bay for the sustained period of winter, during periods of prolonged inactivity the infecting bacterium was allowed to flourish, rapidly gaining steam against the whale's dependent immune system.

Jim shared his evaluations with Lanny, but the ramifications of such a prognosis were immeasurably menacing to everything the project set out to achieve. The idea alone would mandate permanent care. Jim's analysis was scoffed at, ignored. Nothing of Keiko's condition ever went further than the internal communications lobbed between the two continents.

Enter the *Daniel*

Over the first season of escorted walks and introductions the crew and Keiko's time at sea redoubled. The need to keep their charge in close proximity to wild whales often required the walk formation to remain at sea for days on end. Under pressure of foul weather, bucking seas and extended distances from Heimaey, the *Draupnir's* limitations were quickly exposed. She was poorly equipped to offer comfort during drawn-out voyages in the North Atlantic. That had never been part of her criteria in the first place. Now obvious that adventures to sea would involve unknown spans of time, Jeff began lobbying for a more suitable vessel to take the *Draupnir's* place. The search did not take long.

The *Daniel*, a larger and more recent model of a Coast Guard rescue vessel was located on the mainland. OFS leased the vessel that would become Keiko's new walk-boat in the coming season. They did not discard the *Draupnir*, rather, she was demoted to a support role and continued to accompany the small band to sea on the next series of introductions in the beginning of 2001.

Daniel was everything the *Draupnir* was not. Where *Draupnir* was worn, *Daniel* was shiny and new. Where *Draupnir's* lines evinced an older model, *Daniel* carried the sleek thoroughbred lines similar to that of the indefatigable *Thor*. Her cabin gave ample room to a sizable crew, affording them protection from the worst conditions when needed. Her deck space both fore and aft dwarfed that of the *Draupnir*. Piloting the *Daniel*, Michael or Greg would have nearly a 360-degree view through the larger windows wrapping around almost two-thirds of the *Daniel's* pilothouse. Where Jen could scarcely fit herself in *Draupnir's* crow's nest, the *Daniel* comfortably made room for three. The vantage point became a favored spot for Jeff and Blair and the latter often filmed Keiko's introductions from the bird's-eye perch.

For Keiko's benefit, the all-familiar and welcoming platform extended from her starboard side, suspended by lines tied off to the platform's outer corners, the shipside anchored to *Daniel's* sponson. Yet another difference was the presence of convenient handrails that bordered her aft section. The railing proved to be a handy location by which the crew could secure a food bucket, allowing the person working Keiko from the platform easy access without assistance. Every advantage the *Daniel* offered was put to task as voyages to sea began extending well beyond the daily walks of the first season.

Seasons

Summer 2001 began much the same as the previous season ended. Although during the course of this season, the new escort *Daniel* introduced Keiko to wild whales on nearly 100 separate and distinct occasions. The outcome was always unpredictable. The *Daniel's* crew could never anticipate how Keiko or the wild whales would react.

Sometimes Keiko would hastily porpoise away from the wild pod, other times the wild pod porpoised away from him. The interactions were never close enough for physical contact, at least not from the observation standpoint of the walk crew. No close encounter within reach of the pods lasted more than fifteen to twenty seconds before one or another form of erratic retreat took place. Still other times, Keiko quickly moved away from his kind only to stop 200 or 300 meters on the periphery of the pod, floating at the surface, and facing in their direction as if watching, listening. Usually he continued his departure, but always in random directions.

In the first meeting of wild whales, Keiko had bolted from the scene on a north, northeasterly heading. Lanny had proclaimed that Keiko was heading "home." In that season and the one following, it became abundantly clear to the crew that there was no discernible pattern to Keiko's withdrawal. On varied and arbitrary departures from the wild whales, they once found him trailing a cruise ship dumping trash in its wake, and on another he was found alone heading due south. They could never foresee which way he would go or what he might get himself into. On the more convenient outings, Keiko simply returned to the familiar setting of the escorting *Daniel*.

The seasonal presence of other whales around the island chain was finite. They knew that continued returns to Klettsvik Bay only limited Keiko's exposure to wild pods. In the short season not a single opportunity could be missed. At each failure, the solutions stemmed toward logistical improvements. *Surely more time at sea would benefit his tenuous social experience?*

It was in this season of 2001 that the *Gandi* was introduced to the release effort. A 130-foot, nearly 300-ton fishing vessel, the *Gandi* greatly enhanced the team's ability to remain at sea, day and night, under almost any weather pattern.

Utilizing the *Gandi* as a mother ship and the *Draupnir* to ferry crew to and from the base of operations in Heimaey; trips to sea stretched from days into weeks. *Daniel* remained the designated walk-boat, the only boat by which Keiko received direct attention. In company with the *Gandi*, the crew completed multiple trips for sustained periods, the longest of which was forty-five consecutive days at sea.

Gaining comfort in their ability to locate Keiko in the aftermath of ever-unpredictable encounters with wild whales, they often allowed him to venture nearly sixty miles distant before they intervened, seeking him out and recalling him to the *Daniel*. No matter the particulars of each exposure, in every case Keiko ended up back with the walk formation or on his own, apart from whale or human. On the occasions where he left the area of whales and support vessels, his trajectory remained sporadic. On at least one protracted voyage, long separated from the walk crew, Keiko returned to the harbor in Heimaey. But this is as far as he would go. That he did not stay with the wild ones was the only foreseeable ending to each social encounter.

During these adventures on the high seas of the North Atlantic, Keiko's food was limited. What food that was provided by the walk crew amounted to only a scant portion of his normal diet, nothing that would be considered his full requirement. The applied theory: that an increased hunger drive would embolden him to take part in the opportunistic foraging of wild whales. But even those occasions where the crew of the walk formation had witnessed the wild whales engrossed in feeding, Keiko never joined in the activity, merely staying in the distance appearing reluctantly interested in the unfamiliar sights and sounds. At times he exhibited clear signs of hunger evident to the staff that knew him so well. Even still, on the outskirts of a wild pod engaged in feeding behavior, gannets diving the water in pursuit of scraps, Keiko did not take advantage.

In the aftermath of these feeding frenzies, and once the wild pod moved on, Jeff often yearned for Keiko to pick up the scraps left behind, and floating on the surface. On one such occasion Keiko duped the crew into short-lived hopefulness having returned to the *Daniel* with a very small fish in his mouth. Unfortunately, he did not eat the fish; rather, he merely carried the trophy as if to show his shipboard mates. This was the only time Jeff or the walk crew ever witnessed Keiko with fish that they had not provided.

Fashioned from prolonged absences from the walk formation and a minimalist diet, he once went nearly three weeks without receiving sustenance. Jeff knew intuitively that Keiko was not filling the gap elsewhere. Evidence abounded from the simple observation of Keiko's behavior and activity level, increased interest in the *Daniel* and soliciting for his trainer's attentions. More prolonged absences of nourishment produced a stated lethargy. Following the most extreme fasting periods, Keiko would eventually stop swimming altogether, only logging at the surface near the flotilla as if he had lost the energy to move.

Back in the "office" Charles Vinick continued to prop up the operation, and convincing the FWKF board, Humane Society of the United States (HSUS) and even Jean-Michel that Keiko would go free at any time . . . that it was merely a matter of days. A report that Keiko was often seen competing with the wild whales for food was spoon-fed to the media. This version of Keiko perverted actual events taking place around the island chain over 4,000 miles away. Charles assurances—intended to bolster ongoing financial support—when they continually fell short, only augmented the impatience of those bracing up the prolonged release effort.

Onboard the *Daniel* or the *Gandi*, the crew was often accompanied by a variety of videographers pressing for material. Obtaining footage of Keiko's exploits was of paramount importance within the organization, after all, documentaries had been promised. Footage was shot from the helicopter, the *Daniel*, the support boats and nearly every conceivable platform available. Under the dictates of this unrelenting effort, the crew was often required to forcibly lead Keiko, even herding his path, into and among pods of wild orca. Results were so contrived that distaste for the Santa Barbara-based leadership began to take hold among a handful of the more field-tested crew. Still, idyllic clips of the scenes playing out on the high seas would go a long way to garnering much needed financial support back home. But few within the ranks of animal fieldwork have a stomach for such making of the sausage. Indeed, the practice slowly began to erode confidences within the ranks of the release team in Iceland.

On the stage of the North Atlantic, Jeff knew the release was stalled. Unless a change in approach was made, Keiko's moderate interest in the wild whales would go no further. His premonition was only compounded by Keiko's persistent returns to the *Daniel* and stagnant winter's spent in Klettsvik. As fate would have it, an opportunity presented itself, one that Jeff saw as a one-in-a-million chance that they had to take. The very prospect of it drove a wedge in the enduring friendship between Jeff and Jen.

Friendly Whale

The walk formation normally consisted of three key vessels: the walk-boat *Daniel* and the supporting observation boats *Draupnir* and *Heppin*. During the encounters, the formation was often spread out by 1,000 meters and more between boats, frequently out of sight from one another. One typical morning, after escorting Keiko to the vicinity of a wild pod, the walk crews passed the time and like any other day, hoped for a breakthrough. One or two assumed the role of lookout, scanning the horizon and looking for additional whales or other marine species, alternately checking in on the wild pod and visually tracking Keiko's whereabouts.

Onboard other support vessels some listened to hydrophone sounds, straining to hear some form of "conversation" between Keiko and the group of animals. It was amidst this routine roundup of onboard ship activity and small talk when the *Daniel's* radio sparked to life calling Jeff's attention.

"Jeff, Jeff, . . . this is Sammy!"

Jeff responded with passing interest, "Yeah?" Sammy's energy wasn't unusual.

"We have a friendly whale here," Sammy offered excitedly, in his characteristic high-pitched voice smothered in French accent. He asked, "Can I go swimming with 'em?"

Sammy was a photographer hired by Jean-Michel to document the ongoing introductions. He was easily excited and commuted an unwavering fascination with the project and excursions in company with Keiko.

Jeff didn't hesitate. "Sure, go for it," he crackled back across the radio.

In the waters surrounding the *Draupnir*, a young whale had stopped nearby, curiously watching the odd assortment of humans. Sammy slipped into the water, already wearing the customary splash suit worn during ocean walks. Little preparation was required beyond that of zipping up the front of his suit. Jeff had coached Sammy on the proper etiquette at meeting a wild killer whale, chiefly, not to approach the animal but to allow it to approach him.

At Sammy's entrance, the young whale shifted this way and that moving its head as if probing the stranger, but did not leave the scene. Close behind the young whale's assumed mother supervised the interaction. After just a few minutes watching Sammy the mother nervously moved to intervene, swimming between the animated human and her offspring. The two swam to a more comfortable distance keeping a watchful eye in Sammy's direction. The encounter, brief as it was, offered a welcome reprieve from the aboard-ship monotony and generated lively conversation, albeit embellished bantering, on the shared fascination between Sammy and the "friendly whale."

Certainly Sammy's experience that day was exceptional, a story to be cherished as a trophy placed on the mantel of memories. But it was more than that for Jeff, who ceaselessly pondered the unexpected young whale that afternoon and for days following. *What if we can get this whale interested and close to Keiko?* Seemingly, the idea of a whale "meeting us halfway" offered the opportunity of a friend for Keiko, one willing to reach out . . . or at least interested. It was something, maybe only a fingerhold, but something.

Two days later, again Sammy called to Jeff over the boat's radio. "Jeff, Jeff, . . . the little friendly whale is here. Can I go swimming with him?" No one actually knew the sex of the young animal.

Again Jeff encouraged the interaction. Turning to Jen, he said, "Ya know, what we need to do . . . is entice this whale, offer something interesting to look at or watch, I mean we've got a friendly whale, young animal . . . let's encourage that."

Jeff knew that the whale would eventually get bored and they would lose the chance. Jen would have none of it. The very idea of interfering by influencing a wild animal flew in the face of research protocol. But Jeff went further, even suggesting that they offer the young animal a fish.

"Jen, . . . " Jeff pressed, "It's not going to eat frozen herring, you have to teach them to eat dead fish . . . totally different taste. It's more of an olive branch." No matter, Jen was already enraged by the mere suggestion.

The consummate researcher, Jen would have nothing to do with altering the wild whales behavior. Any researcher worth his salt knew that observer bias or worse, influencing the outcome, was the kind of recklessness that would make the Keiko Release Project the scourge of the research community. Jeff's suggestion was a clear violation of the prime directive. So outraged was Jen, that she and Jeff, the parental leadership of the project, did not talk for a week following their confrontation. The overbearing silence between the two was unsettling to the entire team. It was as if Mom and Dad had their first serious fight.

A week after their fallout, Jeff and Jen tired of the conflict. Civility and their longtime friendship allowed the conversation to continue.

"Jen, the reason I got into the Keiko Project was to be able to try to learn something about the wild whales." Jeff instinctively dropped his voice just above that of a whisper but with quiet intensity in his tone. He was very convincing. "If we have a friendly whale, that we can attach cameras to, or ya know, research equipment to that animal and we can get that animal to interact with Keiko . . . wild whales . . . that's what we're here for."

Jen finally capitulated, but with condition. She would agree to continue the occasional swim with the friendly whale, but no offer of herring masquerading as an "olive branch."

That's all Jeff needed. If given a crack in the wall, he could skillfully drive a Mack truck through it.

"So the next time the whale goes out—the next time we get a call that the friendly little whale is around, then you go in the water. I'll let *you* go in the water." His tone had become instructional, allowing a hint of urgency. "You go in, interact with that animal in a positive way. But I'll tell ya, if it was up to me, I would offer that olive branch to that whale."

To Jeff, the herring was nothing more than a recognizable object; he knew the whale wouldn't eat the fish. It was a means to further the relationship that Sammy had started.

Just a handful of days later, "Jeff, Jeff, . . . the friendly whale is here!" Sammy's excitement burst across the radio waves.

Shortly after joining with the *Draupnir*, Jen was in the water. The friendly whale had become more comfortable by now. The interaction was going well. No physical contact took place, although there was much frolicking about the small area, as if two children showing off on the playground without actually playing "together," eyeing each other and trading bodily expressions.

During the odd exchange, the young whale abruptly dove down and disappeared for more than a minute. When it resurfaced, the friendly whale had a stunned herring hanging out of its mouth. Just as quickly, it spit out the herring in Jen's direction, now only inches away.

Standing on the port side of the *Draupnir*, Jeff was incredulous and pointing out his words with his finger toward the whale for emphasis. "Jen, that is *exactly* what I wanted you to do . . . offer the olive branch! That whale is giving *you* the olive branch!"

Despite the fascination displayed by the young whale and shared by the crew, Jen was never truly comfortable with taking the relationship any further. They saw the friendly whale on and off again over the next ten days, but only twice more did they enter the water with the whale. Eventually the little whale stopped showing up altogether. They never offered the herring and never knew what might have happened if they had. During this incredible exchange, Keiko was always in the general vicinity of the walk formation, but seemingly oblivious to what was taking place in the lee of the *Draupnir*.

Gradually and subconsciously, the staff was moved by the spirit of the friendly whale and that which the effort represented. Lacking any form of decisive intent, they began to consider alternatives, other ways outside the proverbial box that might entice Keiko's interest in the whales.

It was on one such occasion that Jim took advantage of circumstance hoping to encourage progress. They had been on the outskirts of a wild pod, watching Keiko watch the whales as they foraged on a biomass of herring. Nothing unusual took place with Keiko during the feeding, he did not engage the pod, nor did he partake of the buffet of herring. As he often did, Keiko merely dawdled on the periphery of the pod within sight of the *Daniel*. Shortly after the pod moved on Jim saw an opportunity. Left in the wake of the feeding frenzy, stunned herring scattered the area and were easily plucked from the surface. In close association with the excited feed and still in proximity of the wild pod, Jim grabbed one of the herring.

Calling Keiko to the side of the *Daniel*, he played with the herring at the water's surface, expecting that Keiko would take the fish, make the connection. Keiko showed little interest in the fish Jim offered, only mouthing the gift and only so long as Jim facilitated the exchange. The moment he left Keiko to his own devices, the herring which hung from his mouth was dropped. Jim tried again, with a second live, but disoriented herring. It didn't happen. Keiko would only passively nudge the fish or momentarily hold it at the outermost extent of his mouth as he had done with the first. He wouldn't eat the proffered food. Jim was befuddled. *Come on buddy, it doesn't get any easier than this,* he thought.

All things considered, this was the same herring that provided Keiko's daily sustenance; the only difference that this herring was fresh, stunned but still alive. Keiko had eaten stunned fish before; in fact it was ever successful in the conditioning trials that involved haddock or cod. *What made the difference now?* Jim wondered.

Stephen Claussen had been watching the attempt. "You'd think he would recognize the herring," he said, assuming it was the type of fish that was the problem.

"I don't know. I doubt it has anything to do with the fish." Jim shrugged as he spoke. "It's almost like he's just timid . . . ya know like when a dog won't take a treat around a more dominant dog."

Jim had an unusual drawl, shaped by a very laid-back, almost Key West-style easiness. His tone was sympathetic.

"He has to be hungry by now. Did you sit him up? Maybe toss it into the back of his throat?" Stephen asked.

"Nah." Jim sighed. With a defeated glance he added, "I asked him to target up, but he didn't respond to my target. Like he's still watching the pod more than me, so I didn't push it." Not much more could be said. Silence filled the air for a few moments. Jim stood where he had been leaning out across the sponson moments before, staring at Keiko who floated head down just a few feet away.

"I just wish he would make the connection . . . here they are feeding and all, same fish and everything . . . and they don't seem to care about him. If we could just get him to take some of the herring maybe the light would go on."

Jim's spoken thoughts summarized what everyone craved, but on the heels of his direct attempts to feed Keiko the stunned fish it seemed there was nothing more clear they could offer the Big Man to get him beyond whatever it was that held him back.

Klettsvik: Take Two

October 2001: Once again winter approached. Night pushed back and day retreated. The seas grew angry. Alongside his escort, Keiko returned to Klettsvik.

Limited financial resources strained the project on every level. Keiko had failed to go free. Another winter of operations in Iceland would have to be endured. What excitement or anticipation that existed at the start of the summer season now faded, leaving little more than administrative frustration behind. Charles undying insistence that Keiko "could go at any time" festered within the halls of the FWKF board. The project was hungrily consuming almost $300,000 a month. They were just coming to terms with yet another holding pattern and six months later, another series of walks. There seemed to be no end in sight.

This, of course, placed the organizations supporting the project in a conundrum. It was the philanthropic euphoria of entrepreneur

Craig McCaw that had given life to the release effort in Iceland. Following an enlightening private in-water interaction with Keiko at Oregon Coast Aquarium, he had promised that the project would not fail for a lack of funds.

Every day has its end.

From the start, it was McCaw's wife who was the true impetus behind his support. Suffering a divorce during the course of the release campaign, ongoing financial support from the McCaw's had continued only by legal requirement; this time with clearly defined limits. He would not abandon the project cold turkey, though pressure on the board to present an exit strategy quickly became all-consuming. At the administrative helm, the task fell heavily on Charles' shoulders.

Like any business under the duress of financial strain, a first survival tactic took aim at slowing the bleeding by cutting costs. Among the trio of FWKF, OFS and HSUS was an organizational culture of proclaimed humanitarianism. By default, and as likely for fear of negative publicity, they refused the cold, hard finality of termination. Rather than fire anyone, they changed the rules of the game. Rotational teams were eliminated. Individual members of the release team would have to agree to live in Heimaey year-round with only one paid trip home each year. Per diem was abolished and housing became the responsibility of the employee. A final blow, salaries were reduced. The measures proved effective.

Over the months following September and lingering onward through the New Year of 2002, one by one the original expat staff rejected the downturn in compensation and living conditions. Jeff and Jen never returned from their winter stateside, unwilling to accept not only the personal impact, but also the drudgery of managing the project on a shoestring budget. Jeff, in particular, was adept at fieldwork; he knew what it meant to work under such strained conditions. In another time and place he might have agreed to the reductions, but Iceland and the Keiko Release Project made the proposition a different beast. It was death by a thousand cuts. They were asking the impossible.

The crew compliment now heavily vested in Icelandic staff. Jim became the soul individual remaining with any experience in animal sciences worthy of note. Those who had previously been in a security role, namely Ingunn, and others who were operations oriented, became the animal specialists charged with Keiko's daily care. It fell to Jim to orchestrate and educate the assorted band of characters. So swiftly had the atmosphere shifted that it warped what were otherwise stable routines, previously made possible by the seasoned cast of expats. Although the original release team had been limited in experience to largely that of the Keiko Release Project, at least they had each been familiar with zoological care. As important, they were well acquainted with existing procedures and protocols, having been amidst the evolution of Keiko's release spanning almost three years of Icelandic operations. The rather blunt transition to a well-meaning, though ill-equipped, staff was a staggering concept. In practical application, it drained Jim in ways he had not anticipated.

At Charles' instruction, Gummi, the ever-loyal and ever-present business manager, sourced additional local experience in an attempt to shore up the operation. That winter, Thorbjorg Kristjansdottir, an educator at a small marine-life facility in Reykjavik was hired on to assist. Called "Tobba" by her closest friends, she constituted the best of what local experience could be had.

Iceland is not home to an extensive and sophisticated marine zoological presence as in the case of the United States. While many Icelanders have been near killer whales throughout their lives, their exposure is that of observers, not deeply involved in the day-to-day care of such animals. This truth is not a slight against the qualities of Icelandic people. Iceland simply did not possess the foundations or institutions that demanded relevant experience vital to a project so highly specialized as the release of a lifelong captive bull killer whale.

By hiring locally, Charles could alleviate the costly expectations of specialized U.S. experience. But in the trade-off, he modified a key element, the very foundation of a project that epitomized the most complex undertaking in the history of marine mammal sciences. Thus, the lofty multimillion dollar world-famous Keiko Release Project was effectively reduced to a shadow of what it once represented.

Tobba was unwittingly launched into a position well beyond her professional understanding and skill level. Any normal human being would jump at the opportunity of a lifetime such as it was, and Tobba reacted no differently. On the surface, she conquered the basic daily tasks required of her and she did so with dependability and commitment. What Tobba lacked in experience, she made up for in her affection, eventually love, for Keiko.

Jim and Tobba worked together through early winter at the end of 2001. However, the downsizing of the project's backbone of experience eventually bore too deep for Jim to reconcile. Though he made attempt, his insistence at bringing back at least a portion of the original crew fell on deaf ears. No matter what the reasoning, nothing turned the tide of shrinking financial resources.

Jim began to see the administrative decisions taking recognizable form in the field. Weighted against the absolute stagnation he had witnessed in the prior season of introductions, Jim sensed the project was heading for a brick wall. Over the months that he toiled, it was Keiko alone that laid like a heavy mist over what was otherwise resounding clarity. Like so many before him, his conscience ultimately demanded his departure from the project, though not withstanding every attempt to convince himself otherwise. It was the hardest decision Jim had ever faced. Unlike his colleagues, his departure represented the final exodus of experienced management. The decision cut deeply into any chance Keiko had at survival. It was no longer just a question of success at freedom, it was now a question of basic existence, most especially during the tenure of human custody.

On the day of Jim's departure, May 5, 2002, the weather grounded the commuter plane, forcing him to endure the four hour ferry ride to the mainland. On its course to the open ocean, the Eimskip ferry passed Klettsvik Bay granting Jim his last vision of Keiko, thus bringing a remarkable chapter of his life to an unceremonious close.

Excerpt—Jim Horton's personal journal
May 5, 2002

The ferry is moving now and I look with admiration at all of the old fishing boats and ships, some ancient and rusting, some tied up alongside each other cramming for space, some still loaded to the gunnels with last night's catch, setting so low in the water they look like submarines. Ah, the stories they could tell, braving some of the worst seas in the world. I pass the cliffs leading to the bay pen, now covered in lush green moss and suddenly I forget all of the reasons for my leaving and am in awe at the beauty while hundreds of sea birds fly about, crying out as if to say goodbye. I climb up to the top of the ferry and stare out at the bay pen and look for Keiko, I think back to all the early morning shows I put on for all of the other staff who were passing by on the ferry for their very last time, getting their last glimpse of Keiko, waving teary eyed, and then they were gone. No morning show for me as I slowly cruise by, Keiko is very sick, again, but this time the worst ever, having spent the last four days just floating at the barrier net, He may very well be on his way out this time and I carry his blood back with me in hopes of finding the problem. I see Ingunn on the roof of the dry house, the little hut on the pen that I had spent so many hours riding out hurricane force winds. She waves and I wave back and once again the butterflies in my stomach begin to flutter about. I have grown to love Ingunn, a single mother of three, petite, yet tough as nails, she started out as night security and then became the only Icelandic killer whale trainer in the world, I leave Keiko in her hands, she being the only one left now that has a relationship with and the rare experience of being the only one remaining that Keiko trusts. I didn't see Keiko, which was probably better anyway. I didn't say goodbye to him either, for some reason it always seems harder to say goodbye to animals than friends, perhaps because they cannot speak the comforting banter of well-wishing and it'll be all rights that always makes saying goodbye to someone just a little easier.

Cobbler, no further than the sandal

In the aftermath of Jim's departure, the project's executive management recognized at once the cataclysmic gap in experience left behind. Charles had been deeply ingrained in the frontline of release operations since shortly after Keiko's arrival in Iceland. He understood better than most the downward spiral in the quality of hands-on leadership. The dumbing-down of the project was palpable to him. He had lived it, he had orchestrated it. Charles had been devoutly loyal to OFS founder Jean-Michel for many years. And though his desire to see the project forth in excellence was in many ways genuine, financial battle lines had been drawn. He now focused on the survival of OFS and protecting his dear friend's name.

Ultimately, OFS relinquished control of the project to HSUS. For OFS the transition was a fire escape. Financially the weakest among the trio of organizations invested in the project, OFS could not sustain the cost of leadership any longer. Effectively, Charles' management of field operations was callously handed over to the supervision of Dr. Naomi Rose, a lead biologist for HSUS. The wealthiest of animal rights organizations HSUS had been financially involved in the FWKF since before Keiko's move to Iceland.

Naomi was notoriously outspoken. Interminably confident in her perceptions, made righteous by what ostensibly rivaled that of a holy mission, her views and her tactics were known within the zoological community. Contrary to the stereotypes of the animal rights continuum, Naomi looked like a fourth-grade teacher, who had just walked off the canvas of a Norman Rockwell schoolyard painting. Her fair-skinned complexion was outlined by bobbing dark hair with graying highlights. Short and a bit plump, she was not physically adapted to fieldwork nor did she instill confidence in those who worked around her with any regularity.

Some would say that in her view of the world, zoos and aquaria represented no more than vile trafficking in animals, an assured genocide carried out by evil perpetrators remiss of conscience. If there was a means to her end, her convictions afforded her great latitude in whatever it took to reach that end. By her actions and her words, she seemed to hold killer whales in higher regard than most other forms of life, animal or human. While the study of killer whale natural behavior was her claim to career fame, her Ph.D. was based on a hundred or so hours observing killer whales in the wild. Naomi had no more experience in the daily care of a killer whale than does a visitor to a SeaWorld park.

Trainers were of no value to Naomi. She believed release was a biologist's domain, and this rightfully and fittingly hers. From the perch in her watchtower, her view of the project was unwavering. She would show them all it could be done, that she could release even the most difficult candidate. She would show them. She would win. Come hell or high water, Keiko would be deemed freed. In Naomi's world, no other outcome existed.

In prior years, Naomi had been no more than a casual guest to the Icelandic base of operations. Her cursory visits to Heimaey in the course of Keiko's rehabilitation had been infrequent. Although she masterminded operations going forward in a more active role, Naomi was not the boots on the ground field leader. To fulfill this vacancy of seasoned leadership, HSUS hired former whale trainer Colin Baird (no relation to Robin Baird). A Canadian in his thirties, Colin's experience with killer whales stemmed from a decade working as an animal trainer at Sealand of the Pacific in Victoria, British Columbia.

Sealand of the Pacific, an antiquated facility, was a remnant of a 1970s ideal in marine life presentation. The small facility was literally a semi-open watercraft no larger than Keiko's bay pen. Only it was not solely limited to housing Sealand's three killer whales, their marquee attraction. In fact, the floating facility moored in a Victoria harbor shared its space with a number of sea lions, a seal exhibit and guest entrance/gift shop. Killer whale housing within the wooden vessel was limited to two pools. The main a rectangular pool contained by a double-walled seine net pursed at depth, the other a small medical pool scarcely large enough for an adult whale to turn about.

Categorically outdated, Sealand was long overdue for its place in the history books. But it wasn't the size of Sealand or its age that were most notable in the context of Colin's resume and newly acquired responsibility over Keiko. What seeped through the cracks of Sealand's history was the arcane training practices employed on and around her decks. In an ironic twist of fate, little difference existed between Colin Baird's alma mater and the Mexican home where Keiko's journey began.

Sealand's animal trainers were people, and like most people, good natured and genuine in their affections for the animals in their care. But donning the hood does not a monk make. Like many facilities before her, Sealand's methodologies were passed down through the ages, their origins steeped in the trial and error of early pioneers in animal training. They did not get into the water with their killer whales. They could not. A product of deprivation, social isolation and punishment, the whales with which Colin cut his teeth in animal training were lost to them through the mix of pseudo-science techniques applied; methods disguised in the old-school language of the trade.

This description is not in its entirety a fair evaluation of the whole. There is little doubt that caregivers at Sealand loved their animals and committed themselves to providing the best care with the tools they inherited. Nonetheless, an acute shortfall of many facilities like Sealand presented supreme consequence in Keiko's plight; the lack of a sophisticated and empirical understanding of the principles of learning.

In the annals of zoological history, in-water interactions with the world's top predator were the impetus behind the advancement of applied behaviorism. So impactful was the early development of in-water interactions with five-ton predators that the very foundations of animal training were erased, only to be rewritten in the manifest of applied science. These advancements in animal behavior and learning far surpassed the one-dimensional practices first employed in marine mammal care. Today, they form the backbone of zoological sciences in the specialized care of countless species. As profoundly important as this understanding is, it cannot be understated as related to leadership of the release effort.

Defying this critical foundation, Keiko's release moved forward with what would equate to third-string experience piloting a jet at twice the speed of sound. On the grandest stage, amidst a delicate phase of the introduction, HSUS gambled with mankind's most ambitious animal reintroduction and the fate of a world-famous whale.

Tainted by the restrictive dictates of budget, their decisions now laid waste to every fragile accomplishment that hung in the balance. This period marked the greatest of many monumental shifts in the course of the Keiko Release Project. Profound undercurrents yet to be revealed began their trajectory here, in the aftershock of HSUS assumed custody over the project. An unspeakable alternative belayed by the unyielding agenda of Naomi and the naïveté of her field commander, Keiko was silently stripped of choice.

Opposing Forces

In earlier seasons, despite other varied alterations in strategy, the staff had maintained the devised neutral position of the walk-boat when Keiko was in the vicinity of wild whales. As agonizingly painful as it was to wait sometimes hours for the whales to clear, it was a prime directive not to interfere. Many of the staff outside the behavior team failed to understand the crucial significance of this directive. Even the starting of the engines or slight movement of the boat at the wrong time could and would influence Keiko's behavior. His history and bond with the walk-boat was unavoidably made stronger still when it became the only source of familiarity in his strange new world. This, compounded by his first traumatic encounter in the wild, made the walk-boat such a powerful stimulus, it became both the path and the barrier to his freedom. We continued to closely follow his doomed saga through communication with our contacts, who provided detailed accounts.

During the third and final season of walks and introductions hosted from the Icelandic base of operations a fatal flaw in judgment betrayed the very goals of the hard-won ocean walks and each chance encounter with wild orca. In the emerging season of 2002, decisions made on the high seas surrounding Vestmannaeyjar became practical, dictated by hands-on people, albeit ill-equipped to understand the roads traveled and the necessary path ahead. Among them Michael Parks, big of heart and unquestionably devoted to Keiko's success, was a boat captain by trade. The staff considerably reduced, Michael now played a substantial role in deciding the daily routine between Keiko and the wild pods. Initially hired on as a laborer in the building of the bay pen, Michael became the only common thread with more time on location than any other single individual.

In Michael's world, things got done by action, not sitting idly by and hoping for the best. To him, the inaction of drifting in neutrality while Keiko floated nearby at the surface yet away from the wild pods was nothing if not an asinine waste of time. *Why sit here and allow him to float while the wild pod disappears over the horizon? We need to get him back to the whales, as many times as it takes.* These thoughts dominated Michael's disposition toward the task at hand. The more he could get Keiko close to the wild pods the better. There was no sense in sitting idle or waiting, it was that simple. *After all, waiting for what?*

As a result of the seemingly innocuous adjustment in protocol, Keiko amassed more time in the company of wild whales than he had in all prior seasons combined. Yet those unions were always facilitated and prolonged by the overbearing edicts of the walk-boat. Each time Keiko departed from the wild whales, no time was wasted; the walk-boat promptly guided Keiko back to the company of his would-be acquaintances. Outwardly, their approach appeared to shift the balance. Keiko began spending more time away from the walk-boat in the wake of each shepherded introduction. In reality, the delicate balance of learning betrayed what the eyes could measure.

Immediately returning Keiko to the wild whales each time he decided to leave them set together a series of incessant consequences—consequences that lent to Keiko's avoidance of the walk-boat.

In the presence of the wild whales, Keiko learned (was taught) that rejoining the walk-boat resulted in a mandated return to the pod. The conflicting elements at work set the conditions for learned helplessness, a trait long dominant in Keiko's disposition. So it was that he neither joined the wild whales, nor returned to his mother ship. Without knowledge of the invisible forces at work, Michael's call to action taught Keiko to place himself safely in no-man's land, in the purgatory between acceptance and avoidance. Keiko became an interloper; a lone whale denied his station, denied a welcoming home.

Throughout July, efforts to integrate Keiko with any available wild pod continued in haste. Following many of the encounters, the team sought to confirm that Keiko was eating something, anything . . . aside from the minor amounts of fish they supplemented.

On two such occasions, they witnessed Keiko diving near a wild pod that appeared to be actively feeding. In the one instance, he was immediately surrounded by five of the larger whales. Keiko quickly swam away from the intimidation. Trained in voluntary gastric sampling, they tested Keiko in hopes that his stomach contents would reveal ingested fish. The results showed nothing.

In early August, he had moved away from the known location where schools of herring occupied the trophic region near the water's surface. Tracking information placed him on top of a biomass of blue whiting and squid. But both species remain deep during the day only to ascend within Keiko's reach at nighttime. Dive data from the sat-tag showed that Keiko had not gone deep enough to reach either food source.

Try as they might, every telltale sign that led to tentative optimism was dashed upon further investigation. Although Keiko was provided food from the release team, the sparse amounts were not nearly enough to sustain a whale of his size. They did not know when hunger would motivate Keiko to find his own sustenance. They also didn't know how long a whale could go without food.

During periods of prolonged absence from the walk-boat the team relied on a special-purpose tracking vessel to keep tabs on Keiko's location. In July, Keiko repeatedly approached the tracking boat, at times remaining close to her side for hours. The crew was instructed to go below decks and ignore the solicitations. The instruction was meant to deter Keiko's interest in the distantly familiar boat. In at least one instance, while the crew waited in

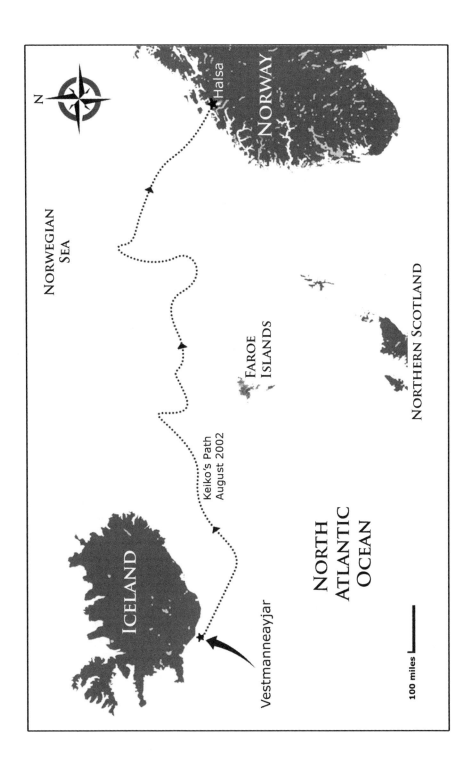

the cabin for Keiko to leave, they could readily hear the sounds of his vocalizations as he pressed his head to the side of the hull.

In each case Keiko eventually gave up and left the vicinity. He never spent more than two hours at the side of the tertiary boat. In contrast, the walk-boat held a much greater power over the whale. July 8, 2002, the walk-boat was drifting idle more than two nautical miles from Keiko's location. The whale had been away from the formation for more than a day. When they started the engines, Keiko headed directly for their position. The prolonged attempt to disregard his presence was fruitless. Keiko remained determinedly close to the boat for nearly two and a half days.

Beyond Faroe

On one ominous occasion in August of 2002, the crew and Keiko had been at sea for an extended period, rehearsing the determined volley between the walk-boat and wild killer whales. Commonplace by this time, Keiko was escorted to a wild pod, the walk-boat initially remained on the periphery, then slowly dropped away from Keiko and the pod. As he had done many times that season, Keiko lingered close to the wild pod, until eventually he and the wild ones were visually lost to the crew of the *Daniel*. They expected Keiko to drift away from the pod after some hours as he always did.

As they waited, an uninvited guest changed the seascape and demanded a shift in their well-worn routine. In the distance, a wall of white water formed along the horizon appearing as if cotton decoration. The torrent of curling waves became apparent contrasted against the backdrop of an ill-omened black sky. The storm would be upon them in a matter of minutes. The crew and the formation, rigged for running, made a beeline for the harbor in Heimaey, almost sixty nautical miles away. They did not have time to find Keiko, and even if they did, they could not afford the crawling three- to five-knot pace he would require.

As fortune would have it, the sailboat *Vamos* was part of the configuration that day. *Vamos* was a third-party sailboat Michael had contracted to provide long-distance tracking. She boasted the experience of seasoned hands that could endure through almost any conditions at sea. Her captain and his seaworthy crew would remain, weather the storm and keep track of Keiko.

Over more than ninety nautical miles, the crew of the *Vamos* tracked the intermittent signal from Keiko's sat-tag. For some time the signal seemed to indicate Keiko was following the wild pod, though no visual confirmation was possible from the low deck of the *Vamos*. Ultimately, they lost contact approximately 150 nautical miles north northeast of Iceland, in the direction of the Faroe Islands, still some 200 or more nautical miles in the distance. Whether Keiko was with the whales or not was impossible to know, but regardless, he was clearly outpacing the *Vamos*, limited by her need to beat back and forth across and into the wind in pursuit.

Empty Nest

Back on dry land in the small town of Heimaey, Michael kept track of the reports emanating from the tracking vessel *Vamos*. But those updates did not last long; very quickly Michael had no more than the once-a-day satellite data to rely upon. Day in and day out, he pinned the waypoints on a marine chart spanning the northernmost extents of the North Atlantic, bordering on the Norwegian Sea and North Sea.

Depicted on the pinned chart, Keiko's movements made the North Atlantic look like a pond, so small it seemed in relation to the distances traveled. But unlike the scattered nondirectional waypoints of the past seasons in company with the walk formation, these new data points sent down from the Argus satellite followed a pattern. Keiko's path actually led north of the Faroe Islands and onward toward the west coast of Norway. Though the path fell short of a straight line, it was nonetheless a steady easterly heading.

On two separate occasions, the tracking effort was supported by aerial survey, a small fixed-wing aircraft flown at low altitude. Once the sat-tag download was received and the waypoints were evaluated, the tracking crew flew to the most recent Argus coordinates in an effort to confirm Keiko's position and status. Regrettably, downloading data only once every twenty-four hours meant that current data was at least three hours old before the tracking team could physically reach the designated coordinates. On both occasions, wild whales were spotted, but only with some measure of searching as they were not at the exact latitude and longitude recorded by the sat-tag. Further, no sighting of Keiko could be confirmed. As conducive as the plane was to covering the distances, it was not the ideal platform for detailed observation. Even the bright yellow of the sat-tag mounted on Keiko's dorsal fin evaded their inspection. Keiko effectively vanished; evidence of his condition could only be extrapolated from pins on a chart.

Lost in translation, HSUS openly promoted Keiko's extended absence as a successful release. Behind the scenes, well removed from the mystery unfolding on the high seas of the North Atlantic, others knew innately that Keiko's advancing absence placed him in severe jeopardy. In fact, the very circumstance at hand had been prophesied by the original release team. Across varied time zones, the original cast watched with growing apprehension as news of Keiko's departure spread.

Well aware of the inner workings of the downsized and defunct release effort, they fretted that HSUS' refusal to acknowledge the broken release, carried out under the worsening conditions of isolation, would push Keiko too far. The insulting fanfare taking stage in the vast abyss of real-time media coverage was enough to stir them to take action. In a diplomatic effort, a sizeable faction of the original release team appealed directly to U.S. authorities. They hoped that knowledge of Keiko's recent history might lead to enforcement of the release permit itself, which, specifically, called for immediate intervention.

татo: Dr. Robert Matlin
Marine Mammal Commission
4340 East-West Highway, Suite 905
Bethesda, MS 20814
VIA FACSIMILE: 301-504-0099

FROM: *Jeff Foster, Stephen Claussen, Jim Horton, Tracy Karmuza, Brian O'Neill, Greg Schorr, Jennifer Schorr, Steve Sinelli*

DATE: *August 31, 2002*

RE: *Concerns related to status of Keiko Project*

We are writing to raise some questions and concerns related to the current status of the killer whale Keiko and the continuing reintroduction effort by Ocean Futures Society and the Humane Society of the United States. Each of us worked for several years on the Keiko Project, including managing animal care during the first two seasons of reintroduction efforts. We are intimately familiar with the past behavior of the animal; the entire reintroduction history; behavior of free-ranging animals in the area; and environmental conditions of the North Atlantic. We are expressing our concerns at this time solely with the goal of ensuring the short and long-term welfare of Keiko. We fully support continued reintroduction efforts, with the caveat that the process should be conducted in a responsible manner with the best interests of the animal as the foremost concern.

We must preface our comments with the statement that our information about the status of the animal is limited, as we have been informed that current staff in Iceland has been instructed by management not to provide any information to past staff members. This has made obtaining current and accurate information difficult and raises concerns that despite our extensive knowledge of the project we are lacking key information about the actual current situation. We certainly hope that our concerns are unfounded and the progress by Keiko towards reintroduction, recently described in media reports, exceeds expectations and that he thrives independently. However, we feel that more extensive documentation of his recent behavior is required in order to determine the success of the project.

It has recently been brought to or attention that Keiko has left the Vestmannaeyjar area and Icelandic waters and has traveled more that 300 miles towards the Faroe Islands, where he is currently somewhere offshore. He is being tracked via the satellite tag and attempts are being make to locate him using the VHF tag. Apparently the VHF signal has been located aerially but no visual observations have been conducted, and no other free-ranging killer whales were sighted in the area of the signal. The distance from Vestmannaeyjar and potential solitude of the animal represents a serious concern for the ongoing reintroduction effort and the long-term safety of the animal, due to a variety of logistical challenges, personnel issues, and potentially regulatory issues since the animal is no longer in Icelandic waters. In our opinion, the ability to intervene using the "walk" boat and return Keiko to the bay pen enclosure has been critical in the past. In order to ensure the safety and well being of the animal, we feel it is necessary to closely monitor his behavior and have the ability to intervene if necessary.

Our primary concern is Keiko's past failure to demonstrate an ability to forage independently in the wild. Although he was trained to eat live fish in the bay prior to reintroduction efforts, this behavior was conducted when humans provided live fish for him; there was no evidence of Keiko foraging independently in the bay. During the 2000 and 2001 reintroduction seasons we saw no indications that Keiko was foraging while in open water, either in conjunction with free-ranging killer whales or independently. This includes periods of up to 10 days while he was independent from the "walk" boat and not provided with food but was closely monitored using a helicopter and tracking vessel. While it is possible that foraging occurred and was not observed, based on extensive surface observations it is unlikely. In addition, diving data collected during 2000 and 2001 did not indicate diving at depths comparable to foraging free-ranging killer whales. Keiko also lost weight during the reintroduction period as demonstrated by body measurements in 2001. In short, as of the end of the 2001 reintroduction season none of us felt that the animal was prepared to survive in dependently without supplemental food.

It is our understanding that earlier in the 2002 reintroduction season, stomach samples were taken following extended periods of Keiko being on his own or in proximity to other killer whales. These stomach samples apparently failed to demonstrate that he was feeding on his own. In addition, to our knowledge he has not been observed or documented via film foraging with other killer whales. We feel that due to the inability to document independent foraging or foraging with free-ranging killer whales in the past, it is critical to ensure that the animal is receiving adequate supplemental food via intervention if necessary. In short, we believe that the precautionary principle should be applied and if it cannot be proven without a doubt that Keiko is foraging independently, then intervention to provide food should be conducted. We believe it is very possible that the animal has already gone for an extended period without food at this time, and that it is humane and necessary to ensure his caloric needs are met. In addition, we do not believe that the fact he has traveled such a long distance is a reliable indication that he is currently foraging. Killer whales captured in Puget Sound in 1976 (Jeff Foster was involved in this project) were maintained in excess of 60 days with no food intake, despite being offered fish on a regular basis, and showed no outward signs of dehydration or starvation. It is our opinion that it could be approximately another month before Keiko begins to show significant physical and/or behavioral indications of food deprivation and that it is premature to claim his current activity as confirmation that he is foraging independently.

It is also our opinion that based on the experiences of the past two summers of reintroduction efforts, "real time" tracking via the VHF tag is necessary to ensure the safety of the animal. This was clearly demonstrated last August when Keiko was very close to shore and had to be recovered using the large tracking vessel despite protocols to the contrary in order to ensure his safety. Unfortunately, had we been relying solely on the satellite locations at this time, due to the delay of the locations from processing, the animal could very well have ended up on the beach. Over the past two summers there have been numerous instances when we have had to intervene and recover the animal with the "walk" boat, frequently using a helicopter to assist with tracking.

Another concern relates to Keiko's behavior while solitary in open water. During the past two summers, Keiko has failed to demonstrate the ability to navigate the waters around Vestmannaeyjar at any significant distance from the bay pen. When removed from the "walk" boat, Keiko frequently demonstrated milling or non-directional travel behavior and occasionally appeared disoriented. However, on a few occasions in response to interactions with conspecifics, he swam in one direction very rapidly for prolonged periods, away from the other animals but in different directions each time. The helicopter was used for tracking and the "walk" boat was used to intervene and gain behavioral control during these situations, as the feeling was that this was not a positive step towards reintroduction, but perhaps a stressful situation for the animal. Our concern is that he may recently have had such a response to an interaction with conspecifics and may now be disoriented.

Another concern involves Keiko's reaction to boats. On several occasions last summer during the reintroduction effort, Keiko approached vessels other than the designated "walk" vessel. These included other OFS vessels such as the Gandi (the large fishing boat) as well as non-OFS vessels. Specific incidents included the animal approaching and swimming in extremely close proximity to fishing vessels on July 12 and August 15 2001, causing concern that Keiko might become entangled in fishing net. During the incident on August 15 the fishing boat was setting nets and we were unable to contact the boat via VHF radio. On another occasion, Keiko broke off from swimming independently and started following a cruise ship moving through the area. Although Keiko's presence at the boats during those periods was not usually prolonged, it sets precedence for this behavior to approach unknown vessels. If this behavior continues and he is in any way reinforced from vessels (either through primary reinforcement by the public feeding him, or by secondary reinforcement via reaction or interaction of people on the boat), it is quite likely that this problem will increase in magnitude and duration. We especially believe this is a concern if he is not foraging independently.

A related and more serious problem involves Keiko's occasional "bumping" of boats. He contacted both "walk" boats and the Heppni Ein (while being used for transfers to and from the bay pen) many times during 2000 and 2001, very occasionally with some serious force. This included lifting the boat out of the water. On one occasion in 2001 Keiko approached a small fishing boat and may have contacted the boat or nets on some manner, as following encounter he has new cracked teeth and missing skin on his back. There was also an incident reported last summer when Keiko apparently bumped a zodiac and/or exhibited solicitous behavior towards the zodiac while away from the "walk" boat. Another concern is that Keiko is habituated to OFS vessels that have jet drives or prop guards, and could potentially be injured by closely approaching a boat with a propeller. Finally, this type of behavior towards fishing vessels, either in Iceland or the Faroe Islands, may potentially endanger the animal through human actions if he is not closely monitored.

A final concern is related to the longevity and experience of current animal care staff involved in the project. As of May 2002, every original American animal care staff member had left the project and a new staff was hired for the 2002 season. The majority of the authors of this letter took part in the rehabilitation of Keiko in Newport, the transport to Iceland, and two seasons of reintroduction. It is our understanding that at this time there is currently no animal care staff member with long-term experience with this animal and/or familiarity with all aspects of Keiko's rehabilitation, history and behavior on site. We are concerned not with the qualifications of the current staff, simply with the lack of an experienced *long-term perspective related to his behavior during previous reintroduction attempts and in comparison to the behavior of other killer whales. We are concerned that this may hamper the ability of the current animal care staff to provide accurate interpretations of his behavior and ability to continue to thrive in the wild.*

The goal of the project has always been to provide Keiko every opportunity for reintroduction while conducting the project in a responsible manner. As stated in the original Reintroduction Protocols submitted to NMFS and the Icelandic Government on May 19, 2000:

The project has been designed and implemented as an experimental effort to determine the feasibility of reintroducing a long-term solitary captive killer whale to the natural environment. During each phase of the project, the welfare of the animal has been the highest priority and careful behavioral and physical conditioning steps have been taken to ensure the highest chances of success for reintroduction. As the project progresses towards open ocean access and potential reintroduction, the outcome will continue to be based on the welfare of the subject animal, as determined by positive physical and behavioral responses. Plans and protocols to increase chances of success, minimize risk, and conduct the effort in a responsible controlled manner have been developed and will be implemented throughout the reintroduction (pg 3).

In closing, we feel that it is critical to closely monitor the animal via direct observations for an extended period and via remote tracking. We realize that the current location of the animal makes this more challenging, but given our concerns we feel it is even more important since he is in unfamiliar waters. Our hope is that as outlined in the original reintroduction protocols, direct monitoring will occur for a minimum of a month and the goal with the satellite/VHF tag is for remote monitoring for a period of one year. In addition, we feel it is important that Ocean Futures and The Humane Society are prepared to intervene on different levels including recapture if deemed necessary.

Anyone every involved in this project (the Board and management included) have invested enormous amounts of time, work, and emotion in this project, and hope to see it succeed. We raise these issues in the spirit of providing the best possible long-term care for the animal. Our goal is to try to ensure the rest of the project is conducted in a safe, responsible manner that optimizes Keiko's quality of life and minimizes risk to the animal.

Landfall

Back at the Icelandic base of operations, Michael continued to plot the satellite waypoints, day after day updating his maritime chart with more and more pins. Finally, twenty-two days after leaving his walk-boat, the last confirmed visual sighting, Keiko hit landfall near Halsa, Norway. Alone.

The information surrounding Keiko's departure from Iceland and arrival in Halsa remained largely steeped in mystery and speculation. Although it appeared likely he initially followed a wild pod during the stormy departure, he did not remain with the pod. No eyewitness account existed to confirm whether or not Keiko was ever with his own kind during the missing three weeks. Regardless of what transpired in that time, the outcome defied speculation. Keiko returned to what he wanted or needed most.

HSUS was officially at the helm during this last fateful season at sea, and thus Naomi Rose played a pivotal role in decision-making. Her perspective on the release stemmed from a foundation of observing behavior in order to define it. Evident in her outspoken animosity toward animal behaviorists, Naomi obstinately denied the very foundations of behavioral science as artificial, assigning them to the constructs of zoological parks alone. *This isn't a show. We don't need trainers to release a whale, we need biologists.* It was thoughts like these that likely colored Naomi's view of what Keiko needed most.

Offensive as the undervaluing of Keiko's conditioning was, this aspect and other deficiencies were only symptoms of a far more

Machiavellian undercurrent. The secretive and shared mandate between Earth Island Institute and HSUS of "release at all costs" ran counter to the decisions taking place on the high seas. Decisions that dictated Keiko's lifelong devotion to man would prevail despite any contrary notion held by the organizational dynamic duo. Whispered within the halls of the FWKF, the phrase "better dead than fed" aptly described the intent and obsession that paradoxically led to Keiko's absolute dependence on man.

In the Presence of the People

First to reach Keiko's location near the small village of Halsa were Colin and a temporary addition to the release team, Fernando Ugarte. Fernando was there in place of Jen to carry on the collection of identification or ID data pertaining to the wild whales. He had worked alongside Colin with regularity over the past several months of the expeditions at sea.

In their first sighting of Keiko, Friday, August 30, 2002, they merely assessed his condition. In an attempt to avoid reconnecting the broken chain of custody, they first observed from afar, hoping not to attract the pseudo-free whale. But Keiko's interest in boats was too great. Investigating the lingering craft, Keiko eventually spotted Colin on her decks. He waited patiently, as he always did, to be acknowledged. Their presence discovered, Colin and Fernando accepted the momentary defeat and allowed the interaction, continuing with closer inspection of their long-lost friend. After three weeks missing, they did not know what to expect. Outwardly, the whale appeared much as he was when they had last seen him.

Throughout this day, much to their liking, Keiko ignored the boating traffic in the area. On his second day in Norway, satisfied with the circumstances at hand, Colin decided to give Keiko a break, leaving him alone to his own devices. The mistake became apparent soon enough. In their absence, children from the small seaside town swam and played with the whale. Images of the encounter spread like wildfire, and soon the spectacle attracted the masses.

All walks of people poured out from the small village, crowding the bay to witness the world-famous whale. Quickly gaining comfort with the storied celebrity, they boarded vessels from makeshift dinghies to small fishing boats making their way into his watery world. Encounters ranged from pats on the head to full body rubdowns rivaling those afforded by his past family of trainers. The boldest of his endearing visitors entered the water with Keiko on occasion, some even crawling onto his back. Keiko could not have been more entertained, swimming from hand to hand and boat to boat. The unanticipated entourage of fans provided Keiko his warmest welcome to a new home he had experienced in a long time.

To Colin and Fernando, the scene before them was profoundly unsettling. Here was Keiko, the recipient of so much effort over so long a journey full of trial and tribulation, nestled snugly in the calm waters off Halsa and entertained endlessly with human affections. He was not with wild whales. Whatever had brought him across the Atlantic to Halsa, be it wild whales, shipping traffic or prevailing currents . . . none of that mattered now. Keiko was back with his human family and seemed perfectly content to remain just so. Colin and Fernando tried desperately to keep the mass of humanity away from Keiko, but no matter their efforts, the attraction could not be thwarted. Friendly Norwegians plastered themselves around Keiko and he, starved of such attentions for so long, accepted them unconditionally.

The attraction went on for several days. At times Keiko seemed drained from the constancy of undying attention and swam away coming to rest in deeper water, only to find himself surrounded again by troops of Halsa citizens. The townspeople fed him on occasion. Keiko willingly took the offerings, at times swimming among the variety of watercraft with his head above the surface and mouth gaping wide.

In the midst of the frenzied love affair, there was some encouragement to be found. When they could, the staff inspected Keiko's outward condition, taking measurements of his girth at multiple locations. By this metric, Keiko had not lost even a centimeter of girth over the course of his three-week journey across the Atlantic. Anecdotally, the staff hypothesized that he must have eaten on his own. *How else could he have maintained his weight?*

If only it were true. Unbeknownst to the inexperienced staff, measurements alone are not an adequate indicator of weight loss in an animal such as a killer whale. A whale of Keiko's size can easily persist without food for more than six weeks without any measurable change in body condition. Furthermore, the appraisal failed to account that Keiko had been conditioned through similar fasting periods in the prior two seasons of open ocean walks.

At this late hour, reliable evidence of feeding over the period could only be gained from blood samples, ultrasound measurements of his blubber layer, skilled evaluation of his post-nuchal fat stores (the area of his head just behind the blowhole) and muscular condition. But even then, for such information to be useful, other comparable baseline information would have to exist from Keiko's state during normal healthy periods. These metrics were not obtained, nor did the comparative data exist to substantiate the information even if they had been.

The first responsibility at Keiko's reemergence in Norway was to confirm that he was unharmed and otherwise healthy. Direction in assessing Keiko's health fell under the purview of Dr. Cornell. Lanny knew any real evidence would fly in the face of assertions that Keiko had sustained his dietary needs for three weeks at sea. A common thread throughout the battle over the release permit itself, Lanny was skilled at avoiding the hangman's noose. He had learned from decades of practice not to give the hangman a noose in the first place. In fact, one of the first actions taken by the misguided release staff was feeding Keiko, an act akin to driving a truck through a crime scene, thus spoiling any chance of clinical confirmation of foraging or the lack thereof.

Conveniently, the *post hoc* reasoning worked seamlessly into the public relations spin proffered by HSUS. News of Keiko's grand adventure across the North Atlantic reached the United States and with it, the idea that sustained body weight proved Keiko had successfully foraged. Across the States the widespread team of expats that had originated the release effort, now sentenced to the sidelines, watched as events unraveled. To a person they each knew Keiko was in jeopardy. They had predicted this outcome. Gnawing frustration, even outrage, ate away at their patience. Keiko had returned to human dependence and yet the vilest of the agenda-driven organizations dismissed this undeniable truth. Tied hopelessly to their mission, the dramatic retelling convinced an unsuspecting public that Keiko had frolicked at sea for three weeks with his own kind. They painted the event as a milestone in Keiko's progress to freedom.

Apart from the court of public opinion, the permit itself required intervention, which among other things mandated lifelong substantive care. By definition, Keiko had failed to integrate socially, failed to thrive. He had not demonstrated avoidance of humans or man-made things. He had purposefully and predictably chosen the only family he had ever known.

Nonetheless and according to news sources, HSUS would continue to encourage Keiko's journey to freedom. In so doing, avoiding the responsibility (and the cost) demanded by the uncomplicated outcome.

Almost immediately, images of children interacting with Keiko and boat loads of people petting the star of *Free Willy* went viral across the Internet. The outcome never more clear, to even the most distanced reporter, Keiko's homecoming and flopped release was aptly described by numerous news outlets. It seemed undeniable to all that Keiko's release had failed, to all but HSUS, to all but Naomi. Reacting to the blitz of human affection, and to no small degree, Keiko's obvious reciprocity, HSUS sought assistance from the Norwegian government in eliminating the entourage of public fondness running afoul.

Honoring the organization's plea, the king of Norway gracefully requested that his fellow citizens respect the needs of the famous whale, offering that their well-intended friendliness and interest in the star conflicted with his sojourn to freedom. The Norwegian public obediently withdrew, respecting their king's wishes with astonishing solidarity.

For nearly a week the first children had befriended Keiko, opening Pandora's box. Then, in the blink of an eye they were gone. With the free-for-all bustle of boats and people subsided, Colin and Fernando could finally oversee Keiko with some semblance of normalcy. But the happenstance location proved inadequate for many reasons, not the least of which was close proximity to shipping traffic and the buzz of nearby townspeople. HSUS and their staff on-site decided to move Keiko to a location far from prying interests or extracurricular human activity.

Taknes Bay

Penetrating deeper still within the catacomb of a tentacled Norwegian coast, they found the serene cove of Taknes Bay a most suitable venue. Here, their surroundings on land were that of farmland and countryside. A languid bay formed their makeshift whale habitat along its way to a watery cul-de-sac further inland. Trips to and from open water were uncomplicated by interferences. Wild whales were known to frequent the area not far away from Keiko's new home and unlike Iceland, their presence was less seasonal, more constant. The location offered more than a few meritorious features in meeting Keiko's needs. What it lacked a great deal were creature comforts for the scant few staff posted in Taknes to oversee ongoing trials.

Settling into the new accommodations was not without sacrifice. Housing for the staff amounted to a shared farmhouse atop a hill far above the water's edge. By comparison to the Icelandic base of operations, they seemed ejected, cast into a bleak yet beautiful exile. Temperatures initially betrayed them; in Norway, the land cools quicker than the ocean. In the beginning Taknes presented unforgettable vistas set in a deceptively warm climate far above the averages experienced in Iceland. But nothing was easy; soon to be made harder still by the earth's growing distance from the sun.

September ever-so-slowly initiated winter. Winter painted most of the land in the white of a snow-laden paradise. Temperatures closer to year's end dropped into the low twenties Fahrenheit. At times the wind channeling through the articulate landscape could whisk away body heat with great aptitude, but the most penetrating cold emanated from a humid seaside dampness. Steam rising from the bay's surface in the still of frigid mornings often concealed Keiko from view. The vibrant palette of landscape, seascape and changing weather patterns presented them with a vision renewed on each dawning day, no two alike.

As it pertained to Keiko's dependent care, they had no infrastructure, nothing that afforded even the most basic conveniences. At the bottom of their daily trek from the hilltop, the only means by which to physically reach Keiko was tendered by a small floating dock extended outward from an ancient rock jetty dappled in blotches of grass. Behind the makeshift husbandry area sat a barn turned fish house. Everything had to be recreated. Supplies, frozen herring upon which Keiko's very sustenance relied; even the rudimentary needs of the small staff demanded careful planning. A never-ending supply list, rich of both common and uncommon needs, accompanied each tedious trek to the nearby village of Liabøen, the administrative center of Halsa.

Management of the ongoing release effort was entrusted to a band of three consisting of Colin, Tobba and Dane. The team was accompanied at times by a tag-team research pair following the project. After essentials of the land-based operations were established, only two of the small tribe remained in the farmstead on Taknes Bay.

Routines with Keiko freely emerged, loosely resembling those of Klettsvik Bay operations. Exercise consisted of the occasional side breach or bow jump commingled with tail lobs (repeated slapping of his flukes on the water's surface) and spinning fast swims selected from the small assortment of trained behaviors. Walks throughout the maze of fjords were implemented on a somewhat fixed schedule of three a week. Facilitated from both dock and boat, husbandry inspection comprised a static portion of Keiko's human face-time. But lacking even the most basic of facilities, evaluation of Keiko's physiological well-being was almost exclusively limited to dead-reckoning intuited through scattered observation and rudimentary lab sampling.

October, November and then December blended together, distinguished only by dropping temperatures, more variable conditions and ice floes migrating outward from shallower waters. New Year's slipped by in the biting teeth of winter. The year pressed ever forward as attempts to encourage Keiko's departure continued unrewarded. Little defined one day from another, not even the usual and curtailed routines of the staff.

In his new world, aside from the interactions decided by his caregivers, Keiko had little means of stimulation. But he had learned how to garner the attention he desperately needed. During prolonged periods absent human guidance, Keiko often wandered away from the drowsy bay. It was a behavior that gained his trainers attentions, often commanding the presence of an accompanying walk-boat. Sometimes the prompted response involved food.

In the dead of winter, on an evening not unlike many others, their monotony was rudely shattered. The coastline was riddled with encroaching ice floes. This was unfamiliar territory for Keiko. Accepting that the whale was prone to happenstance encounters with such things, Colin kept a more guarded and fretful watch on Keiko's whereabouts. Early in the cycle of nightfall, he watched as Keiko swam off. Any other day there would have been nothing worthy of note in the whale's leisurely departure. However, this time Keiko moved steadily in the direction of an ice-packed branch of the bay. It was a limb of the fjord that had no outlet, a dead end.

Fearing Keiko's inexperience with the ice, the project staff boarded the small single-outboard motor boat and followed in the direction of Keiko's last heading. Darkness takes over more quickly on the water and that evening it had become pitch-black. From the sea level position on the small craft, Colin could not find Keiko. But as they approached the ice pack, relying almost solely on the tracking tag indicating Keiko's direction, it became obvious that Keiko had gone beneath the seemingly impenetrable surface.

Knowing that Keiko was in trouble, they used the tone recall to beckon the whale back to their location and clear of the ice pack. For some time they neither saw nor heard any sign of their charge. Without the benefit of sight amid the penetrating black surrounding them, their sense of hearing became intensely heightened. At long last, Keiko's position was revealed, not by sight, but by the repeated sound of ice impacted, then fracturing.

Keiko had navigated well under the field of ice, some 200 meters from the safety of open water. What they heard were the sounds of a frantic whale breaking through the frozen layer. The escape required no small amount of effort, evidenced by repeated dull strikes and muffled, partial exhales. Keiko was well distanced from the thinned edge. Isolated from clear water and thus life-giving oxygen, he had run out of time. He did the only thing he could do to survive. Keiko beat forcibly with his head and back against the overhead ice until at last he could claim a single crucial breath.

Initially disoriented by the heightened state of anxiety, Keiko largely ignored the first attempts at recall. However, shortly after his breakthrough and during continuous and repeated recalls, Keiko finally found his way clear of the ice. He did not escape the chance encounter unscathed. In fact, the skin on Keiko's back spanning from the top of his rostrum to just forward of his dorsal was devastated. Bloody patches revealed areas where his outer dermal layers were completely flensed or stripped by his frightened and desperate beating against the thicker, jagged ice layer.

Keiko, a sprite twenty-seven years of age, looked as if a ragged ancient beast of an animal, his scars and wounds angrily covering the majority of his head and back. So cracked was the once mirror black surface of his skin that it appeared as if a pot left to burn on the stove, the contents dry and flaking from the bottom. Witness to such external damage was sickening, though little actual threat was posed by the physical injury itself. While wounds of this nature can become infected and abscessed, the clean waters of the Norwegian bay discouraged such complications. Even so, in the days following the episode the staff applied preventative cleansing to forestall the risk of infection.

Considerably more foreboding was the impetus for the behavior. This delicate component, the invisible source of neurosis, was by and large overlooked; or as likely, regarded as a random accidental occurrence having no actuarial relation to Keiko's concealed state of distress. At the heart of the coping behavior that painted Keiko's back in red lurked an underlying frustration worn so deep that yet even more severe ailments were slowly taking irreversible hold, manifesting themselves within.

Inward Appearances

Pressing on, the mission guiding the release team did not deviate from the overbearing aim of freedom at all costs. In many ways, assurance of his eventual freedom became hardened by the tale of Keiko's Atlantic crossing. Retold in growing legend, the fateful three weeks at sea became like folklore in the halls of HSUS. *Surely he had proven his ability to adapt on his own, to survive.* The idea became indisputable; an insistence. Far removed from the ambitions or aspirations of his human caretakers, Keiko languished. Though he would venture from the limited scope of their observations at night, he remained ever faithful in his return, forever wanting of their attentions.

Under the guise of the protracted release plan, Keiko's food was regarded as supplementary. A stubborn continuation of the earlier season among the walk formation, they either believed or hoped that their charge was actively foraging. Somewhere, somehow and beyond their sight, possibly in the night. They fantasized that he was finding nourishment of his own accord. They did not feed him much from their hand.

Imposed by their design, the idea constituted the extent of sophistication in the revised release strategy. They theorized that hunger would motivate his interest in venturing out, finding his own nourishment or perhaps even engaging wild whales in pursuit of food. Upon this logical deduction they gambled everything. In stark contrast, Keiko remained in Taknes, unfaltering, never refusing the meager offerings of his overseers.

His behavior belied their dangerous calculation. Wearing onward toward spring, when not otherwise directed under escort of boat or human affections, Keiko merely floated about Taknes Bay. At times, hours in the shadow of contact from his chosen family, he returned to the violent thrashing behavior witnessed early in his Icelandic indoctrination.

To Keiko, the new world defied recognition. In a moment, he was bathed in playful love. In another moment he was left to his own devices, unpredictable spans of time between. He waited. Always he waited. Eventually they would again acknowledge his presence. In time, they would return. They always returned. He listened. He would meet them at the dock or hear the wanting sound of the small boat being boarded. Maybe he would follow the boat on an adventure. Maybe he would carry them on his back about the bay. He didn't know what would come. He waited. Hunger kept him company. He was always hungry. Tired. Eternally tired.

Summer 2003 bathed the bay in immaculate beauty. Mountains speckled with new growth and a returning vibrancy granted a warming backdrop from the hilltop view over Taknes. Brilliant green grass carpeted the span between the crew dwelling and the immense bay beyond. Life renewed lent an air of optimism to those that would accept it.

Not all did.

Heavier thoughts distanced an otherwise carefree beauty of the northern land. Keiko did not look well. Though he conceded to whatever was asked of him and reliably returned the affections of his trainers, he did very little else but rest, stilled at the surface. Day in and day out, he remained inert, scarcely casting a ripple across the glassy surface surrounding him. Keiko, his choice and his needs, were buried under the wreckage created at the convergence of agenda, negligence and ineptitude.

Free

November 2003. Metabolically Keiko's body deteriorated. So long had he been minimally nourished, his once abundant stores were now hopelessly depleted. Lacking any other source of fuel, his system had turned on itself, extracting what little energy he had left from his own tissue. Remiss of the life-giving water contained in his fish, chronic dehydration had long ago set to purpose wreaking havoc on his vital organs. His old familiar enemy found a home in his lungs and flourished against a vastly weakened immune system. Outwardly, his great size and blanket of blubber subverted attention from the struggle within.

Grasping at his second December in Norway, Keiko spent his time mostly alone in the still waters of the bay. On the human calendar it was Thursday, December 12, 2003; in many ways an idle Thursday. Only Dane and Tobba remained. They were wary.

The two recognized a sharp decline in Keiko's behavior. What before went undetected, hidden by size and the survival's clever guise, became evident seemingly at once. Even in this late hour, his condition was not revealed on physical merit, but what they could observe was not the Keiko they had known. It was enough to stir them to action.

Absent means of any other form of intervention, they stuffed what antibiotics they had on hand into the scant few fish that Keiko would take. But the ailment was too deeply rooted. Worse, Keiko's system was beyond its ability to make use of the artificial support. They might as well have attempted to douse a forest fire with a garden hose. Ostensibly, the staff had done all they could. Evening set upon them. Tomorrow they would continue with more medication.

During the night, Keiko's strength to stay upright and near the surface drained from his body. No longer able to support himself at the surface, he dropped slowly to the darkening depths, finally coming to rest, his massive pecs gently propped on the sea floor. Absolute silence surrounded him. As the cold stark deadness of night turned toward the life of day, his life emptied from his body. Once a great and mighty animal, forceful of breath, his last feeble exhale came unanswered by the all-familiar inhale. At this, the end of his long sojourn to freedom, apart from both human or whale, Keiko died.

"Keiko," his given name, means "Lucky one."

Echoes

December 14, 2003, Robin and I were traveling together to South Florida. As we often did in our makeshift mobile office we talked ad nauseam dissecting e-mails, planning responses and discussing the lofty goals of our business. In the four-hour drive we each became captive audience to the other's every thought. We raised our voices in the sharing of ideas amid the drowning bass drum of the big diesel engine pushing us down the turnpike at seventy miles per hour.

Lost in forming the next series of thoughts, I was scarcely aware that Robin picked up his PalmPilot and checked for new messages. He looked at it only briefly in the respite from conversation, then held the device in his wheel hand for a time, staring at the road ahead. The latency of his reaction and the drawn-out pause caught my attention. A heavy change in atmosphere within the broad cab of the F350 was palpable. *Something's wrong at home* I thought. Without a word, without even looking in my direction, he handed me the Palm. The message was still there on the small screen. Once sentence. "Keiko died in Norway." The message had come from a longtime friend and colleague.

Emotional investment in an animal is an extraordinary thing. When that relationship is forged in the course of supremely challenging circumstances, an entirely new level of bond can be experienced. If it's true that learning that takes place during traumatic episodes becomes all-but hardwired into our psyche, then I have to believe that what we retain from experiences during our most devoted acts must wield a vastly higher order of permanence. I had never freed myself from Keiko's plight. On occasion when the topic entered conversation among colleagues, I would spend several subsequent nights involuntarily grinding my teeth in my sleep.

We, along with other members of the Icelandic team, were well aware of what faced Keiko in Taknes Bay. Every attempt HSUS and FWKF made to contain damaging facts from the public or even private domain was wasted effort. The community of marine mammal professionals is considerably small and fiercely loyal in the care of mutual subjects. Through informal communication within that circle we knew full well that Keiko was destined for a life in human care. By the actions taken over the last year in particular, we also knew that Keiko was hopelessly lost. The news did not come as a surprise, yet it shook us to our core just the same.

Months earlier, dreading exactly the outcome which now glowed on Robin's phone, we sent a formal proposal to the Norwegian government outlining the means by which Keiko could remain in the ocean for the rest of his natural life. He would receive the sustenance and human care he so clearly needed. But the plan had surely fallen on deaf ears. It did not succeed in gaining the right audience. Weighing heavily in the snubbing, we suspected that HSUS was so firmly rooted in the political scene in Norway that our attempts at intervention likely never saw the light of day.

These thoughts did nothing but compound the silence that permeated the cab of Robin's truck. For more than two and a half hours, neither of us spoke. For the longest time, I couldn't look over at Robin, convinced it would crack my stoic hold over the turmoil I locked inside. We had put everything we had and everything we were into Keiko. Meticulously erecting and then dismantling and dissecting every step of the rehabilitation. We framed what had been Keiko's best chance at freedom. Though we struggled at times, the foursome of Robin and I, Jeff and Jen were the right mix. Had we had prevailed through the discord, I can be so bold as to say that Keiko would still have been alive.

Respiratory ailments had long plagued Keiko throughout his time in Iceland. To a person, every soul on the project knew this fact. Had he truly been on his own, he would have died a swift death. That he was pushed beyond all reason and rationale, slowly starved in half-measures, only prolonged his suffering. He died of a very preventable condition. Indeed, the outcome was cultivated through actions imposed by the responsible party ordaining "release at all costs."

Worse still, the practices of his caretakers, steeped in ignorance, amassed great confusion in Keiko. They urged him to be free, to seek out a home and family of his own all the while continually welcoming his presence in Taknes Bay. By their misdirection, they persistently injected the abusive turmoil over and over again. Blanketing him in affection and play, they salted the recipe of malnutrition with a neurosis; unseen but evident just the same. In a language and purity only afforded animals, Keiko made his decision. He never strayed from this resolve, demonstrating his choice of human companionship with resounding clarity time and time again.

The choice was never his in the first place. This truth dominated our thoughts for the rest of that regrettable day and for many weeks and years to come. We had let ourselves believe that ignorance would be outweighed by compassion; that in the end, love of an animal would prevail even against man-made agendas. Literally and symbolically Keiko represented a vast array of personal, social and political ideals spanning both time and geography.

In life, he was undoubtedly the most famous whale in history. In death, he became the most famous case of animal abuse the world could not yet fathom.

Author's Note

Keiko's story teaches us that public awareness and value framework for our oceans and its inhabitants can be imposing forces for good. But there is a vastly more meaningful lesson to be learned. When impassioned people wrap themselves so completely in imagination and emotion, but leave knowledge, experience and critical thinking out of the equation, the outcome is almost always plagued with misfortune. Keiko did not die a justified death, nor did he die at a natural age. He was killed.

The attempt to free Keiko was not an exercise in conservation, it was a reaction born of empathy and derailed by ignorance and agenda. It was not science, it was conscience. This alone is not a tenant of responsible stewardship.

If we truly accept our role of stewardship, if we recognize that every environment is subject to threat imposed by a century of industrial expansion, we begin to comprehend that a monumental undertaking is bearing down on our time.

Release is an option. It is but one tool in a vast array of disciplines required in responsible wildlife management and species preservation. Zoological science is another tool, an increasingly vital one.

Everything I have learned over three decades working in the marine mammal field teaches me there is no separation between zoological and wild as it relates to wildlife management. The two areas of expertise, though seemingly set in opposing plots, are interdependent. Knowledge from each benefits the other. In the modern world, one cannot exist without the other.

Truly effective conservation is born of prosperity. As individuals, we know this to be true on a personal level. We do not pay heed to the needs of our immediate surroundings when we are unable to put food on our tables or keep roofs over our heads. Corporations are no different in this regard. Neither are NGOs.

Likewise, it is well-run and prosperous zoological facilities that not only provide the best expertise and environments in the care of their animals, but also contribute greatly toward sustainable conservation work. They do so not only financially, but also through applied expertise, equipment and labor resources. As importantly, they provide a medium for personal contact and exposure to exotic species most of us would never lay eyes upon by other means. This truism is but one portion of the cornerstone in the modern mission of zoos and aquaria.

As a society, we think symbolically. We find it difficult to explain complex topics such as Christianity, but most understand what is represented by the symbol of a cross. Likewise, individuals don't easily grasp the importance of an amorphous undertaking, such as marine life conservation. But we understand the likeness of a dolphin or a killer whale. We know that we care about these animals. Collectively, involuntarily, these symbols motivate us to protect these creatures and the environments in which they exist. This we understand clearly, in fact, almost subconsciously.

The ocean is the earth's air filter, a massive sponge soaking up the worst of our offenses over a hundred years of industrialization. In the last five to ten years, we've only just begun to see the physical symptoms of persistent ocean contaminants on higher order marine predators. Cases of marine mammals dying in the wild are increasing seemingly daily. It is not a symptom exclusive to endangered species; rather, it is inclusive of all species that rely on the ocean as their home. We know we're in for a fight if we want generations of our grandchildren to know an animal like a killer whale—wild or not.

Zoological expertise is an asset that belongs to all of us. In the future of marine life conservation we will undoubtedly face many more trials. In that future we must rely on the considerable arsenal of knowledge and experience at our disposal if we are to preserve not only a single animal, but an entire species. If there is any hope of sustaining a species and its habitat, implementing effective preservation, it will be born of the absolute union between zoological and wild animal sciences, not their division, and most certainly not from the exclusion of zoological experience. In many ways, Keiko's death was the price paid for refusing this important truth.

Bibliography

Baldwin, John D. and Janice I. *Behavior Principles in Everyday Life*. Prentice Hall, 2001.

Bossart, G. D. "Marine Mammals as Sentinel Species for Oceans and Human Health." *Veterinary Pathology*, 2011.

Bossart, G. D., Cray, C., Solorzano, J. L. Decker, S. J., Cornell, L. H., and Altman, N. H. "Cutaneous papovaviral-like papillomatosis in a killer whale *(Orcinus orca)*. *Marine Mammal Sciences* 12:274-281, 1996.

Brill, R. L.; Friedl, W. A. *Reintroduction to the Wild as an Option for Managing Navy Marine Mammals*. San Diego: Naval Command, Control and Ocean Surveillance Centre/RDTandE Division, 1993.

Ferrick, D.; Wells, R. "Effects of stress on the dolphin immune system." University of California, Davis; Chicago Zoological Society, 1993

Friday, R., Thomas, C., Pearson, J., Peacock, P., Walsh, M. "Appetite Fluctuations in Cetaceans Associated with Non-illness Factors." *IMATA*, 1994

Gales, N. and K. Waples. "The rehabilitation and release of bottlenose dolphins from Atlantis Marine Park." *Western Australia. Aquatic Mammals*, 1993.

Hoelzel, A. R. and G. A. Dover. "Genetic differentiation between sympatric killer whale populations." *Heredity*, 1991.

Kazdin, A. E. *Behavior Modification in Applied Settings* (6th ed.). Belmont, CA: Wadsworth, 2001. (Translated into Spanish and Chinese)

McBain, J.; Smith, A.; Stott, J.; Geracia, J.; Krames, B.; Kohn, B.; Johnson, I.; Ridenour, R. "Summary Report of Evaluation Panel Convened to Assess the Health of Keiko." 1998

Moore, M. "Report to the Free Willy Keiko Foundation." 1998.

Scarpuzzi, M. R., Lacinak, C. T., Turner, T. N., Tompkins, C. D., and Force, D. L. "Decreasing the frequency of behavior through extinction: An application for the training of marine mammals." *IMATA*, 1991.

Simon, M., Hanson, M. B., Murrey, L., Tougaard, J., Ugarte, F. "From captivity to the wild and back: An attempt to release Keiko the killer whale." *Marine Mammal Science*, 2009.

Turner, T., Stafford, G., McHugh, M., Surovik, L., Delgross, F., and Fad, O. 1991. *The effects of Context Shift on behavioral criteria and memory retention in killer whales, Orcinus orca. IMATA*, 1991.

Additional Reading

Bigg, M. A., Ellis, G. M., Ford, J.K.B. and Balcomb, K. C. *Killer whales: A study of their identification, genealogy and natural history in British Columbia and Washington State.* Phantom Press and Publishers Inc. Nanaimo, B.C. 1987.

Cook, M.; Varela, R.; Goldstein, J.; McCulloch, S.; Bossart, G.; Finneran, J.; Houser, D.; Mann, D. *Beaked whale auditory evoked potential hearing measurements.* College of Marine Science, University of South Florida, St. Petersburg, Florida, 2006.

Lyrholm, T., Leatherwood, S., Sigurjonsson J. "Photoidentification of killer whales (Orcinus orca) off Iceland." *Cetology*, 1987.

Simila, T. *Behavioral ecology of killer whales in Northern Norway.* Ph.D. thesis, University of Tromso, Norway. 1997.

Wells, R. S., Bassos-Hull, K., Norris, K. S. "Experimental return to the wild of two bottlenose dolphins." *Marine Mammal Science*, 1998.

Index

A

Adams, Douglas, 67
anchor system, 60–61, 74, 136, 153–154, 164, 187
animal abuse, 13, 295
animal behavior, 37, 70, 73–74, 355
animal trainer types, 73–74
animals in captivity, 13–14, 19–20
audible signals, 65

B

Baird, Colin, 353, 354, 370–371, 374
Baird, Robin, 288, 290, 293, 296–299, 324, 353
Behavior Team, 93, 107–108, 138, 145, 156, 166–169, 185, 190, 206, 209, 215, 220, 233, 299, 331, 356
behavioral modification, 21, 35–43, 66–73, 91–92, 100–109, 144, 160, 168, 174
Bell, Phyllis, 11–12
blood samples, drawing, 102, 146, 204–205, 269–270, 369, 372
boomer ball, 29, 69, 96, 128, 176

C

captivity, 13–14, 19–20
Claussen, Stephen, 63–64, 69–70, 93, 96, 98, 120, 131, 140, 211, 220–226, 277, 280, 346, 363
collecting whales, 25
Cornell, Lanny, 41–42, 109, 146, 159–163, 166, 205, 242–244, 250–

255, 298–299, 302–304, 311–312, 315–316, 372
Cousteau, Jacques, 148–149, 295
Cousteau, Jean-Michel, 41, 148–149, 287, 341–342, 352

D

"Dancing Queen," 101
Daniel, 337–352, 360
Differential Reinforcement of Alternative (DRA) technique, 171, 177, 195, 210, 219, 227
Discriminative Stimulus, 65–66
"doff and don" procedure, 170
dolphins, 20–21, 35–37
Donner, Lauren Shuler, 9
Donner, Richard, 9
Draupnir, 192–197, 212–223, 234–236, 256–257, 263–278, 280–287, 296–306, 308–310, 312–315, 318–325, 328, 331–334, 337–339, 342, 345

E

Earth Island Institute, 12–13, 27, 29, 370
Eyjólfsson, Guðmundur "Gummi," 137–138, 241, 264, 267, 349

F

fish delivery system, 209–211, 220, 277
Foster, Jeff, 40, 50, 53–54, 70, 80, 107, 124, 128–129, 133, 212, 245–247, 253–254, 263, 269, 278–279, 286–292, 298, 312, 331, 334–338, 341–344, 348, 363, 382
Free Willy, 9, 22, 24, 27–29, 38–41, 285, 373
Free Willy/Keiko Foundation (FWKF), 27–29, 38, 41, 76–79, 82, 92–94, 119, 126–129, 159–163, 191, 194, 284, 286, 297, 341, 347–348, 352, 370, 382
Friday, Robin, 31–43, 47–55, 60, 64, 69, 72, 75, 78–82, 90–93, 97, 106, 110, 115–118, 124–130, 137, 142–165, 172–174, 187, 190, 195–205, 224–225, 241–244, 247–282, 286–296, 303–329, 331–336, 381–382

G

Gandi, 339, 341, 366
Garrett, Howard, 78–79, 82

gate conditioning, 166–185
genetic history, 22
Griffin, Edward, 25
Gudrun, 155–156

H

Hallsson, Hallur, 198
hand signals, 65, 101
Hanson, Brad, 244–245
Harðarson, Lina, 135
Harðarson, Smári, 135–136, 138
Heppin, 57, 134, 235–236, 256, 263–266, 277–278, 281, 286, 297, 328, 342
Herring Delivery System, 209–211, 220, 277
Horton, Jim, 335–337, 346–352, 363
How to Train Your Dragon, 23
Humane Society of the United States, 341, 352–354, 362, 368–369, 372–373, 378, 382

I

Icelandic Coast Guard (ICG), 151–152
immune suppression, 10, 27, 91, 380
Ingunn, 138–139, 263, 349, 351
instincts, 284

J

Jean-Michel Cousteau Institute, 41, 287, 341

K

Karmuza, Tracy, 93, 109, 110, 113, 182, 184, 199, 215, 226, 235–236, 258–260, 263, 273, 305, 363
Keiko
    anchor system for pen, 60–61, 74, 136, 153–154, 164, 187
    becoming independent, 226–230
    capture of, 10, 26
    collection of, 26–27
    criteria for release of, 87–97
    death of, 380–381, 384–386

description of, 59
diet of, 56–57, 68–69, 147–148, 165, 340, 358, 372, 378–380
disposition of, 71
gate conditioning for, 166–185
ice floes and, 376–378
illness of, 9, 27, 145–148, 204–206, 336–337, 351, 379–380, 383
introductions for, 294–328, 331–332, 338–346
lessons of, 13–14, 384–386
medical pool for, 49, 98, 166–185, 196–200, 247–250
meeting, 9, 12, 45, 59–64
meeting whales, 294–328, 331–332, 338–346, 358–362
net construction for, 125–126, 163–170, 187–188
odd behavior of, 64–65, 224
recovering, 315–325
releasing, 163–164, 196–198, 331–355
setback for, 315–325
social behavior of, 90–91, 284
tracking, 308–315, 358, *359*, 360–365, 376–378
tracking tag for, 244–252, 280, 294–298, 314–315, 333, 358–365, 368, 375–377
Keiko Foundation, 27–29, 38, 41, 76–79, 82, 92–94, 119, 126–129, 159–163, 191, 194, 284, 286, 297, 341, 347–348, 352, 370, 382
Keiko Release Project, 15, 28, 33, 36–40, 51–53, 59, 72, 80, 85–86, 109, 140, 158, 198, 239–242, 271, 331–356, 399
killer whales
  breeding of, 19–20
  in captivity, 13–14, 19–20
  collecting, 25
  communicating with, 21–23
  environment of, 21–23
  eyesight of, 67, 88
  food for, 23
  genetic history of, 22
  hearing abilities of, 67, 88
  knowledge of, 26

motivating, 23
as predators, 24–25, 67
relationship with, 21–23
respect for, 26
social behavior of, 19–24
sonar abilities of, 67, 88
training, 21–24, 34–35, 65–66, 73–74
traits of, 24, 67
in wild settings, 19–20
Klettsvik Bay, 29, 40, 49, 56–61, 89, 121, 125, 131, 140–145, 153–154, 157, 164, 172, 186–207, 229, 240, 270, 276, 316, 325, 334–335, 339, 351, 375
Klettsvik Bay, Iceland, 29
Kristjansdottir, Thorbjorg "Tobba," 349, 350, 375, 380

L

Lacinak, Thad, 67
Lanterna, 233, 244
live-fish conditioning, 190–192

M

Marine Mammal Protection Act, 28
Marine Operations and Research, 93–94, 126, 152–154, 158, 163–166, 169, 186–188, 211
Marineland, 26–27
Mate, Bruce, 11
Matlin, Robert, 362
McCaw, Craig, 348
McRea, Karen, 70, 93, 98, 102–103, 128–129
medical pool, 49, 98, 166–185, 196–200, 247–250
Mexico City, 9–13, 26, 28
Moby Doll, 24

N

Namu, the Killer Whale, 25
National Marine Fisheries Service (NMFS), 28, 244
net construction, 125–126, 163–170, 187–188

Newman, John Henry Cardinal, 19
Newport Coast Aquarium, 12
Niagara Falls Aquarium, 26–27
Noah, Peter, 40, 50, 70, 106–107

O

Ocean Futures Society (OFS), 41, 43, 50–51, 76–83, 90–93, 126–129, 164, 168, 171, 239–241, 287–288, 295–297, 337, 348, 352, 366–368
O'Neill, Brian, 93, 109–112, 168–169, 215–216, 226, 247, 258, 261, 363
Orcinus orca, 26
Oregon Coast Aquarium, 12–13, 28, 71, 89, 348
Orozco, Jose, 11

P

papillomavirus, 9, 27, 145
Parks, Michael, 152, 163, 164, 169, 186, 216–219, 230, 235–237, 263, 269–270, 272–278, 280, 302–304, 308–315, 320–323, 337, 356–357, 360–361, 369
pen anchor system, 60–61, 74, 136, 153–154, 187
pen gateway, 166–185
Phillips, David, 12–13
puffins, 121–123

R

radio tag, 244–245, 280, 314–315, 368. *See also* tracking tag
Reed, Kelly, 156–159, 173, 175, 328–329, 331
Reino Aventura Amusement Park, 9–12, 26–28
"Reintroduction Protocols," 26, 28, 159–160, 368
relationship with whales, 21–23
release criteria, 87–90
release outline, 83
release plan, 87–116
release prerequisites, 87, 124
Release Project, 15, 28, 33, 36–40, 51–53, 59, 72, 80, 85–86, 109, 140, 158, 198, 239–242, 271, 331–356, 399

release steps, 93–97
release team, 45–86
release team appeal letter, 362–369
Richards, Dane, 257–258, 261, 375, 380
Rivera, Diego, 11
Rose, Naomi, 352–354, 369

S

Sanders, Tom, 173–185, 194–195, 198–201, 247, 259, 261, 295–299, 303, 328–331
satellite tag, 244–246, 294–298, 333, 358–365, 368. *See also* tracking tag
Scarback, 11–12
Schorr, Greg, 53–55, 186, 194, 211, 218, 220–223, 226, 337
Schorr, Jen, 50, 53–55, 241, 263, 272, 274, 278–279, 286–293, 331, 334–335, 338, 343–344, 348, 363, 382
Sealand, 353–354
Seattle Public Aquarium, 25
SeaWorld of California, 109
SeaWorld of Florida, 34–35, 156–157, 173, 225, 231
SeaWorld of Ohio, 156, 174
Sili, 57, 213, 268, 271, 279–281, 297–298
Simmons, Alyssa, 80–81, 117–118, 156, 231–241
Sinelli, Steve, 70, 93, 102–105, 158, 211, 278, 363
Siqueiros, David, 11
Siti, Captain, 138–139, 263, 296–300, 302–303, 323
social behavior, 90–91, 284

T

Taknes Bay, 374–378, 382
Tilikum, 231–235
tracking Keiko, 308–315, 358, *359*, 360–365, 376–378
tracking tag, 244–252, 280, 294–298, 314–315, 333, 358–365, 368, 375–377
training whales, 21–24, 34–35, 65–66, 73–74
Turner, Ted, 156–157

U

Ugarte, Fernando, 370–371, 374
U.S. National Marine Fisheries Service, 28, 244
U.S. Navy, 87

V

Vamos, 360–361
Vancouver Aquarium, 24
Vestmannaeyjar project, 28, *30*, 32, 43–48, *44*, 86, 107, 124, 139, 143, 149, 153, 164, 174, 231, 239, 263, 294, 308–312, 331, 356, 364–366
Viking II, 286, 295–306, 323, 331
Vikingur, 297
Vinick, Charles, 50, 72, 77–80, 94, 124, 128, 159–162, 172, 225, 239, 244, 253, 263, 269, 286–292, 300, 311–312, 325–329, 341, 349–352

W

Warner Brothers, 9, 27
Woods Hole Oceanographic Institute, 126, 163, 187, 214
Wyland, 9, 14

Z

Zero-Nine-Zulu, 296, 298, 308–314
zoological science, 355, 384–385
zoological settings, 19–22, 26, 385

About the Author

Mark A. Simmons grew up in a Northern Virginia farm community where his father taught him at an early age that stewardship of wild and domestic animals requires lifelong responsibility and commitment. At age eighteen, Simmons had his first encounter with whales and dolphins at SeaWorld in Orlando, which quickly led him to a ten-year career in animal behavior there, working mainly with killer whales.

In 1987 Simmons began his career in behavioral sciences working almost exclusively with killer whales. In 1998 he formed a consulting firm, Wildlife International Network along with highly regarded marine mammal expert and close friend Robin Friday. The following April, Simmons joined the Keiko Release Project as the director of animal husbandry and led the behavior team on-site in Iceland. There he authored and applied the behavioral rehabilitation blueprint for reintroduction that gained approval from the Icelandic Ministry of Fisheries for Keiko's formal release.

Simmons went on to create Ocean Embassy, whereby his team assists governments on protective marine legislation, participates in ongoing research with marine mammals and is heavily involved in the rescue and rehabilitation of sick and stranded animals. He has provided consulting on marine mammal health assessment and recovery, training program evaluation and development, and zoological program management to numerous agencies such as NOAA/NMFS and facilities worldwide in United States, Mexico,

Singapore, Bahamas, Dubai, Philippines, Iceland, Jamaica, Panama, China and St. Lucia.

Simmons also created and continues to provide visionary leadership of a large-scale research and conservation database called OERCA that serves global wildlife management needs. He has taught marine mammal behavioral science at the University of Miami and conducted numerous seminars and public lectures on the Keiko Release Project.